THE DEVELOPMENT AND PROPERTIES OF RAW COTTON

AGENTS

AMERICA	THE MACMILLAN COMPANY 64 & 66 FIFTH AVENUE, NEW YORK
AUSTRALASIA . . .	OXFORD UNIVERSITY PRESS 205 FLINDERS LANE, MELBOURNE
CANADA	THE MACMILLAN COMPANY OF CANADA, LTD. ST. MARTIN'S HOUSE, 70 BOND STREET, TORONTO
INDIA	MACMILLAN & COMPANY, LTD. MACMILLAN BUILDING, BOMBAY 309 BOW BAZAAR STREET, CALCUTTA

PLATE I.—THE FIRST PURE COTTON.

This photograph shows the first commercial sample of pure-strain cotton passing through a full-sized power-gin, at the Giza Cotton Experiment Station (Egyptian Government). *December 30, 1913.*

THE DEVELOPMENT AND PROPERTIES OF RAW COTTON

BY

W. LAWRENCE BALLS, M.A. (CANTAB.)

LATE FELLOW OF ST. JOHN'S COLLEGE, CAMBRIDGE
FORMERLY BOTANIST TO THE KHEDIVIAL AGRICULTURAL SOCIETY OF CAIRO, AND TO THE EGYPTIAN GOVERNMENT AGRICULTURAL DEPARTMENT

WITH SIXTEEN FULL-PAGE ILLUSTRATIONS
AND TWENTY DIAGRAMS IN THE TEXT

A. & C. BLACK, LTD.
4, 5 & 6 SOHO SQUARE, LONDON, W.
1915

THE SEAL OF THE OLD KHEDIVIAL AGRICULTURAL SOCIETY CONSTITUTES THE COVER-DESIGN, BY PERMISSION, IN TRIBUTE TO FORMER COLLEAGUES.

PREFACE

THE simple story of a cotton-fibre has been made difficult in the past, because of a general disinclination to recognize the law-abiding habits of plants, and it has not been made easier by the tendency to regard cotton as a special gift of Nature, destined to man's use. Further, our knowledge of cotton has lagged far behind our knowledge of other useful plants, because it was but rarely that the trained student found himself living in the cotton areas, and it was still less often that—so living—he possessed the necessary tools wherewith to exercise his craft; lastly—having the tools—he ran the risk of obsession by the financial significance of the cotton-plant

The main purpose of this book is to present the history of the development of cotton-lint, for although this development is essentially normal and simple, it may possibly be of some ultimate use that the mystery which has enwrapped it should be removed.

Accessory to this purpose it has been needful to indicate the manner of the development of the plant on which this lint is borne. In doing this I have taken occasion to point out some of the more recent views and methods which the "organized common sense" of natural science has brought to bear on cotton, and also to indicate the practical bearings of such views and methods.

It cannot be denied that the latter aim of this book

is the more difficult. Practice draws average lines of conduct through the medley of practical considerations, and except in the cultivation of Pure Strains, and the shortening of the Picking Intervals—both of which are too expensive to employ except on high-priced cottons—the researches have resulted in little of immediate applicability. I have endeavoured, however, to leave the matter in such a form as will enable the results of future scientific researches on other plants, and on animals also, to be fitted to the special case of the cotton-plant with as little waste of time and trouble as may be.

My greatest difficulty has been due to the very limited appeal which my subject makes to a very wide audience, whom it is nevertheless desirable to reach. The possibility of a purely popular treatment in this book was rejected as too remote, besides being dangerous with a relatively unfinished topic. On the other hand, all technicalities and jargon outside those pertaining to cotton have been deleted wherever it was practicable to do so. Some care has been taken to facilitate perusal by the employment of varied type, and by the use of marginal notes indicating the main interest of paragraphs as relating to the seed, growing, irrigation, ginning, grading, and spinning of cotton.

Comparatively few references are made to the writings of previous authors, and this has been done deliberately, because the subject has suffered severely from injudicious copying of accepted statements without verification. A list of the chief works read is appended.

Three causes have led to this comparative independence of treatment. In the first place should be set the researches and influence of Mr. F. F. Blackman, Reader in Botany in Cambridge University, which have revolutionized our knowledge of the workings of plants, and reconstituted the available data. The author was one of those who first applied Mr. Blackman's methods to the study of Growth, first of a fungus under the microscope, and then tentatively to cotton-plants growing in the open field.

Secondly, a statistical repetition of O'Neill's work on the breaking strain of cotton lint hairs, made by Mr. F. Hughes, Chemist to the Egyptian Ministry of Agriculture, was of very great use. Mr. Hughes ascertained the precise significance of such results, and showed from this that an unexpectedly small number of hairs was sufficient to give useful figures. It having thus been shown that single-fibre testing was worth doing, an automatic machine for doing it was the natural outcome, and we now know not only the value of such tests, but also their useless features.

Thirdly, my own system of routine records of field crop, in the form of Plant-Development Curves, accumulated from 1904 to 1914, has provided abundant material from which check data could be drawn as occasion arose.

The scope of this volume may be criticized as embodying much that should properly find its place in scientific journals only. It must not be forgotten, however, that the greater number of its probable readers have not the opportunity to consult reference libraries, although the subject of "Cotton" in all its manifestations is far more of an entity to them than that of botany. That this association should have been injurious to research in the past is not a valid reason for withholding information from the growers and users of the future cotton crops.

W. LAWRENCE BALLS.

LITTLE SHELFORD,
CAMBRIDGE.
January 1, 1915.

CONTENTS

CHAPTER I

THE DEVELOPMENT OF PEDIGREE

CHAPTER II

THE DEVELOPMENT OF THE PLANT

CHAPTER III

THE DEVELOPMENT OF THE BOLL: I. STRUCTURAL

PAGES

CHAPTER IV

THE DEVELOPMENT OF THE BOLL: II. ENVIRONMENTAL INFLUENCES

CHAPTER V

THE DEVELOPMENT OF COMMERCIAL LINT

CHAPTER VI

THE DEVELOPMENT OF COTTON-GROWING

APPENDIX I

METHODS OF INVESTIGATION

APPENDIX II

TABLES OF STATISTICAL DATA

LIST OF ILLUSTRATIONS

FULL-PAGE PLATES PRINTED SEPARATELY

PRINTED IN THE TEXT

THE DEVELOPMENT AND PROPERTIES OF RAW COTTON

CHAPTER I

THE DEVELOPMENT OF PEDIGREE

THE reader of a book dealing with the various kinds of raw cotton, or with the cotton trade, will notice a very large number of tolerably confusing names—Oomrawattees, Uplands, Islands, Nyasaland, and so forth. If he is engaged in the manufacturing side of the trade, many of these will be familiar to him, but possibly without any conception of the different kinds of plant on which they are borne. If the grower's side of the trade is his affair, a few of them will be very familiar ; but he might even fail to recognize plants of unfamiliar kinds for cotton-plants at all.

Reference to works dealing with the botany of cotton may easily bewilder any reader of an inquiring turn of mind. Either the multiplicity of names leaves him with the impression that these things are better left to botanists, or on further inquiry he finds that different names are given to the same kind of cotton. In point of fact, the main outlines of the Systematic Botany of the cultivated cottons are relatively simple, though the details are almost incapable of resolution.

Object of Classification.

The object of all classificatory, or systematic, biology was, in the first instance, to provide a designation for every known organism by which it could be conveniently mentioned without circumlocution. At a later date the idea of a common ancestry for those forms which were closely similar took root and grew until the old purpose of mere convenience was overshadowed by the purpose of tracing relationships. For many years after the "Origin of Species" was published this newer purpose was productive mainly of argument, but the present century has seen a revival of experiment in this direction, with consequent advances in knowledge.

Genealogical Tree.

The end and aim of such inquiry is thus the construction of a genealogical tree which shall show the evolution of each organism from extinct or surviving ancestors. Such a tree has the advantage of being pictorial, therefore easily memorized, and serving a more definite purpose than the assignment of names which—to the trade, at least—are merely useless duplicates of easier names.

The genealogy of any cultivated crop is necessarily intricate, owing to transport of seed from one country to another, and to natural or artificial crossing of stocks thus obtained, with the consequent formation of commercial varieties which embody not only the original wild varieties, but many compounds of elements inherited fractionally from them.

PLATE II.—SENNAAR TREE-COTTON.

One-twentieth natural size. For description see Sir G. Watt's "Wild and Cultivated Cottons."

is to sail dangerously near controversy, but the fate of its descendants is less uncertain.

The other characters of this primitive cotton were those common to related genera within the suborder *Gossypiæ* (or *Hibiscæ*) of the order *Malvacæ*, which is very definitely distinct from the other suborders, and includes such obviously cotton-like plants as *Hibiscus* and *Abutilon*. All have capsule fruits, as distinguished from the one-seeded fruitlets of the Mallows, and, in common with all the *Malvacæ*, they possess a staminal column, formed by the united development of many stamens into a tube which surrounds the style, and bears a brush of stamens externally.

At the present day certain wild cottons are found which represent the descendants of this primitive ancestor not so very much altered, such as the wild species *Gossypium sturtii* in Australia. The existence of this latter form indicates that the genus was definitely cotton-like and probably widely spread before the Australian continent was isolated from Eurasia, so that cotton is by no means a new genus, even from a geological standpoint.

The modifications which led to the cultivated cottons of to-day may be sketched as follows:

Evolution of the Various Species of Cotton.

At an early stage the offspring of some primitive cotton-plants threw off a group having leaves cut into rounded lobes, and of only moderate size. From this stock are descended the plants which we may group for convenience as the "Asiatic cottons," comprising the majority of cultivated Indian cottons, Levant cottons with the extinct cotton of mediæval Northern Egypt,

This intricacy is shown by cotton at least as much as by any other crop, but only in details. The main outlines are quite simple in so far as the commercial cottons are concerned, especially since some of the most obvious differences are really of little importance; thus, although Tree cotton and Annual cotton would appear to be very primary divisions for the genus, the actual differences are but slight: a difference of a few degrees in the relation of growth to temperature, the constitutional power to develop one lateral bud instead of another, or even a change of district only, and the annual becomes a tree, or conversely. Similarly, the smoothness or "fuzziness" of the seed, which has been ridden to death in some schemes of classification, is almost an accident; various forms of the accidental result happen to be commoner in some species than in others, but naked-seeded forms are known now in all the commercial cottons, having probably arisen as sudden "sports."

Classificatory Characters.

In constructing a genealogical tree, we are compelled to take some account of its trunk, or, for our purpose, of the primitive cottons; but to do more than glance at them would carry us into regions of controversy. The original ancestor of cotton was probably a hairy annual plant, with rounded leaves, yellow flowers blotched with crimson and surrounded by three green leaves, ripening a fruit divided into five or more compartments, each containing seeds covered with a green felt. Even to postulate such an ancestor

Origin of Cotton.

and some indigenous African tree cottons (Fig. 1). The evolution of "lint" had possibly taken place before this group-parent was thrown off.

Probably much later in the world's history the primitive stock threw off another modification, in which the leaves were larger and more or less cut into pointed lobes, the blotch of crimson on each flower was rather smaller, and the whole plant was less wiry than in the Asiatic group.

FIG. 1.—COTTON LEAVES. (VERY DIAGRAMMATIC.)

To illustrate the general differences in form and size between the three main groups of cotton-plants. Above, Asiatic; left, Upland; right, Peruvian.

While forms which may well represent the primitive ancestor of the Asiatic group are still surviving, no apparent representative of this next offshoot is known. In all probability it did not long exist as a separate form, but underwent another pedigree-cleavage, into plants with more deeply cut leaves which retained the yellow flower, and plants with the less cut leaves which lost the yellow flower colour (Fig. 1 and Pl. III.). The descendants of the two branches of this cleavage have given rise at the present day to the "Peruvian group" and the "Uplands group" respectively. The former embraces the Sea Islands, Peruvian, and Egyptian cottons; while the latter is commonly typified by American Uplands, Cambodia, and the Hindi Weed cotton of Egypt. The origin of the former group was probably in Central America, while Persia or China is indicated as the original habitat of the latter. It should be remembered that the form from which we have designated the latter group is entitled to the honour only on account of its commercial importance, having been imported to America from Asia.

Three Main Branches.

It would be interesting to attempt to follow the subdivisions of the genealogical tree through the ages, but such discussion would be nine-tenths pure speculation, eked out by fragments of evidence from dried specimens, from the atlas, and from the beginnings of experimental work on heredity in cotton, by which the inherited structural components are slowly being analyzed out and traced to their source. For our present purpose it will suffice to leave the matter at the simple conception of three main branches (Pl. IV.), with a few slender twigs coming off at intervals to hint to us what the nature of the extinct ancestors might have been.

We may next discuss the origin of the lint itself (Pl. V.), and of its accompanying "fuzz."

The most primitive of the surviving cottons vary in their seed coatings from a single coat of fuzz to fully

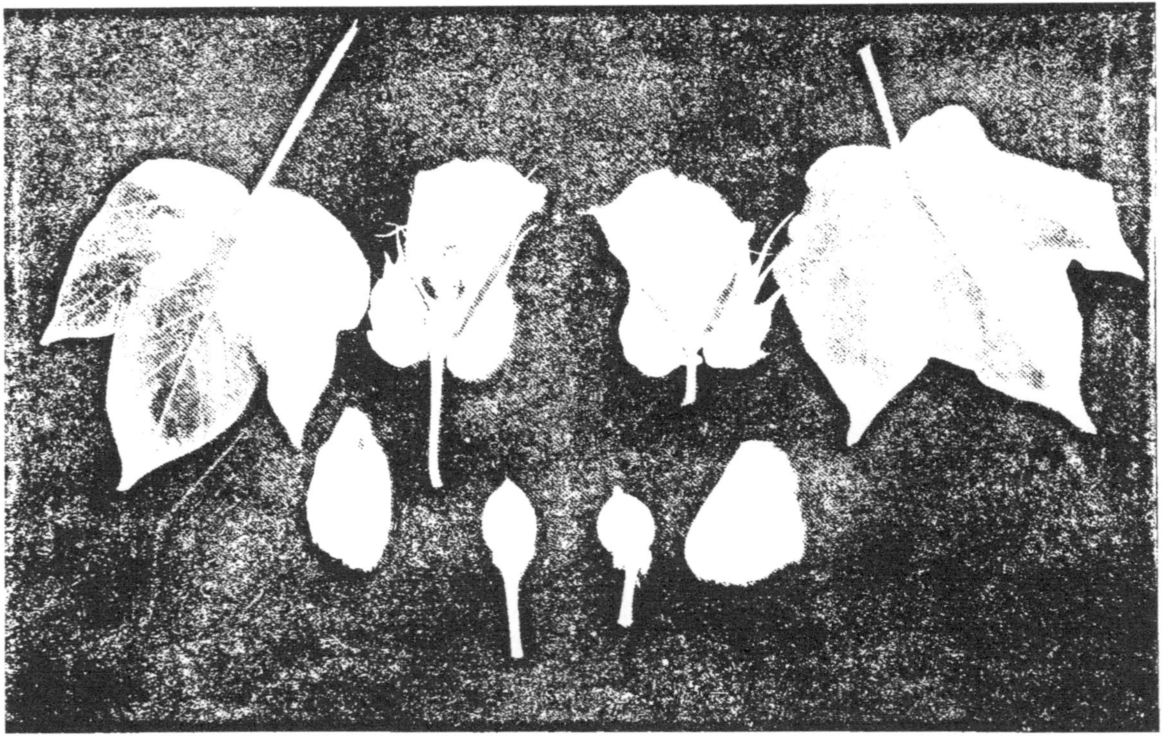

PLATE III.—LEAF, FLOWER, PETAL, AND BOLL.

differentiated fuzz and lint. Whether the modern lint is the primitive fuzz enlarged, with a new kind of fuzz below it, or whether the lint is a new development above the primitive fuzz, is not easily ascertainable, and it is, indeed, quite possible that evolution may have taken place in both ways. That there is very little essential difference between fuzz and lint is quite certain, and in any case the original evolution must have taken place by the formation of a new layer of seed hairs. Both lint and fuzz exhibit similar colourings, due to closely similar—if not identical—chemical substances, through greens and browns to white. It may not be generally known that cotton with lint of a vivid emerald green is sometimes found in American Upland, and is known as "Texas wool." The behaviour of these colours on crossing is the same in lint and fuzz, as also is the distribution of lint and fuzz on the seed.

Evolution of Lint.

From the primitive cotton-seed, with its coating of fuzz, there thus evolved a seed with two coatings, and it may be of interest to consider the effect of such evolution upon the chance of the plant in the struggle for existence.

Utility of the Lint.

It has been repeatedly asserted that the lint is an adaptation for wind distribution, but the probability of this statement is very dubious. That open bolls may be stripped of their cotton by the wind is undeniable, but there the matter ends, short of a cyclone. The seeds are not blown out one by one as a rule, but in a lock of six or seven; they are too

Wind Distribution.

heavy in proportion to their hair surface to travel more than a foot or two before reaching the ground, and unless the surface on which they fall is very clean and tidy—which it is not under jungle or meadow conditions—the lint rather obstructs than facilitates the further transport of the seed, through entangling by its movements (as it dries or moistens) the projecting portions of any plant or object on which it lies. Moreover, the matted fibres hinder the field germination of the seed, and it is easy to recognize self-sown seedlings which have grown from fallen seed cotton by their emaciated appearance as compared with seedlings from ginned seed. In point of fact, the modern cultivated cottons do not stand the slightest chance under wild conditions in competition with the meanest weeds, although their lint is developed to what should be a highly beneficial degree. On the other hand, though such locks of seed germinate badly, they rarely fail to germinate, owing to the moisture-absorbing properties of the blanket of lint; and it is certain that a less development of lint, insufficient to retard the germinated seedling, but still sufficient to retain plenty of moisture, would give such seed a very good chance of survival in competition with naked seeds under conditions where rainfall was intermittent or scanty. The natural habitat of cotton would thus appear to be land with ample water beneath the surface, which its long tap-root could ultimately reach, but with scanty rainfall. On such sites the cotton-plant would possess decided advantages over many others. In any case, it is time that the cherished fiction of wind dispersal was abandoned.

The subsequent evolution of the fuzz and lint which all the three main groups of the genus possessed from the commencement is best sketched in terms of unit-factor composition, as ascertained by the application of Mendel's law.*

Evolution of Fuzz.

The fuzz in the Asiatic cottons appears to depend on a single factor, which may be lost, and naked-seeded sports or varieties then appear. In the Peruvian and Upland groups there are certainly two factors concerned at least, the loss of one of them producing a seed with fuzz almost entirely confined to the two ends of the seed, and the loss of the other or of both producing an entirely naked seed. The appearance of naked-seeded forms, such as Hindi Weed, within the Upland group, and possibly of naked seeds within Upland varieties themselves, would seem to be due to the modern loss of one factor; while the typical naked or semi-naked seed of the Peruvian group, which is older than history, seems to be due to the loss of the other factor far back in evolutionary history. How far this generalization may go can only be settled by much laborious accumulation of data from the study of hybrids, but it is certainly true in some cases, such as a first cross of naked Hindi Weed with the semi-naked Egyptian, which is covered with fuzz like Uplands, and behaves in later generations in such a way as to show clearly that two factorial elements are involved.

The history of the fuzz seems thus to be one of analysis, the full fuzz of the primitive cotton being progressively split up into simpler forms by the loss of factors. That of the lint is certainly the reverse, new forms appearing by synthetic evolution,† first bearing lint in place of no lint at all, and then

* A general outline of this subject is given in "Mendelism," by Professor R. C. Punnett. London, 1911.

† The appearance of lint, or an increase in its length, may be interpreted as analytical evolution, if the author's views on growth-inhibition are substantiated.

long lint in place of short lint. Similarly, perhaps, the variations in distribution on the seed may have arisen, but more probably the primitive lint originated all over the seed coat, and has become irregular in its distribution by loss of factors, just in the same way as the fuzz, down to sports which are found in American Upland and in Hindi Weed, producing naked seed with neither fuzz nor lint! Similarly, there are indications that mutations may take place in long-linted cottons at the present day, whereby short ancestral lint reappears; but the experimental difficulties in keeping a cotton-plant's pedigree untarnished are so great that it must be many years before any definite statements can be made on this subject

Effect of One Characteristic on Another.

It may be considered somewhat absurd to state that the lint length in every variety of cotton depends on inherited factors, in that it would seem to demand an endless number of factors, or at least one for every eighth of an inch increase in length. As a matter of fact, no such demand is made; some three lengths at most would cover the whole range of raw cottons, the gradations being provided by a process for which the author has devised the term "Autogenous Fluctuation," as distinct from ordinary Fluctuation due to external circumstances; in this process the manifestation of a character is affected by the inherited nature of the plant body on which it is borne. Thus, if a $1\frac{1}{4}$-inch lint borne on a medium-sized seed is transferred by crossing to a large-seeded plant, it will rise in length to about $1\frac{1}{2}$ inches; and, conversely, if placed in a small-seeded plant, it will fall to 1 inch. Many other similar effects can be traced, due to the inherited size of the boll, the leaf area, branching, and other

PLATE IV.—PLANTS OF THE THREE MAIN GROUPS.

Left: Asiatic type (Saharanpur country cotton). Centre: Upland type (Truitt Big Boll). Right: Peruvian type (Egyptian Abbassi). Photographs taken at the beginning of the flowering period. One-fifteenth natural size.

more recondite peculiarities. It is to this phenomenon, superadded to a very few constitutional changes in length, that the whole range of length in different kinds of cotton is is due.

Fluctuation.

To the ordinary Fluctuation (Pl. V.), which acts on the constitutional basis just described, producing differences between the crops in different parts of the same country and in different years, we shall advert when discussing the development of the fibre in the principal portion of this book.

Commercial Varieties.

The subdivision of the three great divisions of the genus *Gossypium* into those ultimate units in which the crop is classified on Cotton Exchanges may be carried to very fine distinctions of breed, or it may be very rough, according to the social conditions of the country of growth.

As a rule, even in the least fine cottons, the trade name covers a population of plants which are fairly closely related, forming a "subspecies" of the genus, though a different name may be given in the trade to the produce of the same subspecies when grown in another district. In any serious cultivation, however, and especially in the finer cottons, the subspecies is cut up into named varieties, which have usually originated in the chance discovery of some well-favoured plant, and the multiplication of its descendants.

The discovery of such especially good plants implies that the subspecies itself is not homogeneous, since we know that if the abnormality had been simply due to accidents of nutrition it would not have reproduced itself. A few remarks on the material from which commercial varieties are thus derived may not be out of place.

Subspecies.

We have seen that all cotton-plants can be classified on broad evolutionary grounds into three main species, and several equally important, though economically useless, minor primitive species. We have abstained from carrying the classification farther on account of the absence of experimental evidence, but the next step would be the grouping of all species of cottons of the Peruvian type, for example, into groups of relations, each group being designated a subspecies. The members of a subspecies would all be alike to ordinary observation, just as all the brambles in a hedgerow are obviously brambles. Closer observation would reveal differences other than those due to accident of situation and nutrition, whose nature could be tested by raising offspring from self-fertilized seed.

Elementary Species.

If this test be applied to subspecies of cotton, it is found that many separate components which breed true at once go to make up the subspecies, in addition to a larger number of plants which do not breed true. These latter are necessarily of hybrid origin, though the cross which originated them may have taken place even centuries before; the former may or may not have originated from a cross. Where the circumstances are such as to justify the presumption that they did not originate by crossing—and such circumstances are rare in cultivated cottons—these definitely distinct forms are classified as "elementary species," just as any gatherer of hedgerow blackberries will have noticed that two bushes growing side by side may have slight but definite differences; the English bramble, in point of fact, can be subdivided into a large number of such elementary species.

Where a cultivated subspecies of cotton consists of more than one elementary species, the first step towards its improvement is the separation of these elementary species from one another, and their cultivation under distinct names.

Improvement of Cultivated Cottons.

Short of improvements in site, in water-supply, and in cultivation, this step is also the last one possible, short of resort to artificial hybridizing. An elementary species is as definite a thing as a chemical compound, and may, indeed, be regarded as such. It cannot deteriorate nor improve its constitution, however much may be done to improve its environment, nor is it of the slightest use to select within it for the best-looking plants.

In the commercial cottons, however, no original elementary species can yet be traced, although investigations into the genetics of cotton are showing the various components of the cultivated stocks, and may ultimately permit us to declare how those components were combined to form the original elementary species from which the stocks arose. Thus, in the case of Egyptian cotton, we know that Sea Island and an indigenous brown cotton of similar habit, closely resembling modern Peruvian, were the original

Extinct Elementary Species.

components; we may safely presume that both these components had originally consisted of several elementary species, so that the pedigree of a modern variety of Egyptian cotton is a very complex one.

The origin of these elementary species has taken place in the same way as for the subspecies and species—namely, by abnormal germ-cell formation. The modern view tends more and more towards a physico-chemical conception of living organisms, and in the case of species formation it is being more generally accepted that a new species arises from its parent species at a single jump. Instead of forming its germ cells by symmetrical cell division, so that the offspring resulting from reunion of male and female cells exactly resemble the parent, something goes wrong with the physico-chemical machinery of cell division, and abnormal asymmetrical pairs of germ cells are formed, with the result that, on fusing with one of the opposite sex, a representative of a new species, subspecies, or elementary species, is produced. The process of sudden origination of new forms in this way is called "mutation." The proof of its occurrence demands most careful experimentation, and, as we mentioned formerly, it will be years before such proof can be obtained clearly in the particular case of the cotton-plant, though it may well be still taking place.

Natural Crossing.

Except for the purpose of clarifying ideas upon the subject, it is of little immediate use to discuss elementary species in the cultivated cottons, since none are recognizable. This is due to the fact that free intercrossing takes place under natural conditions between related forms of cotton. The Indian group does not appear to cross with the Upland or Peruvian groups, but the two latter can easily

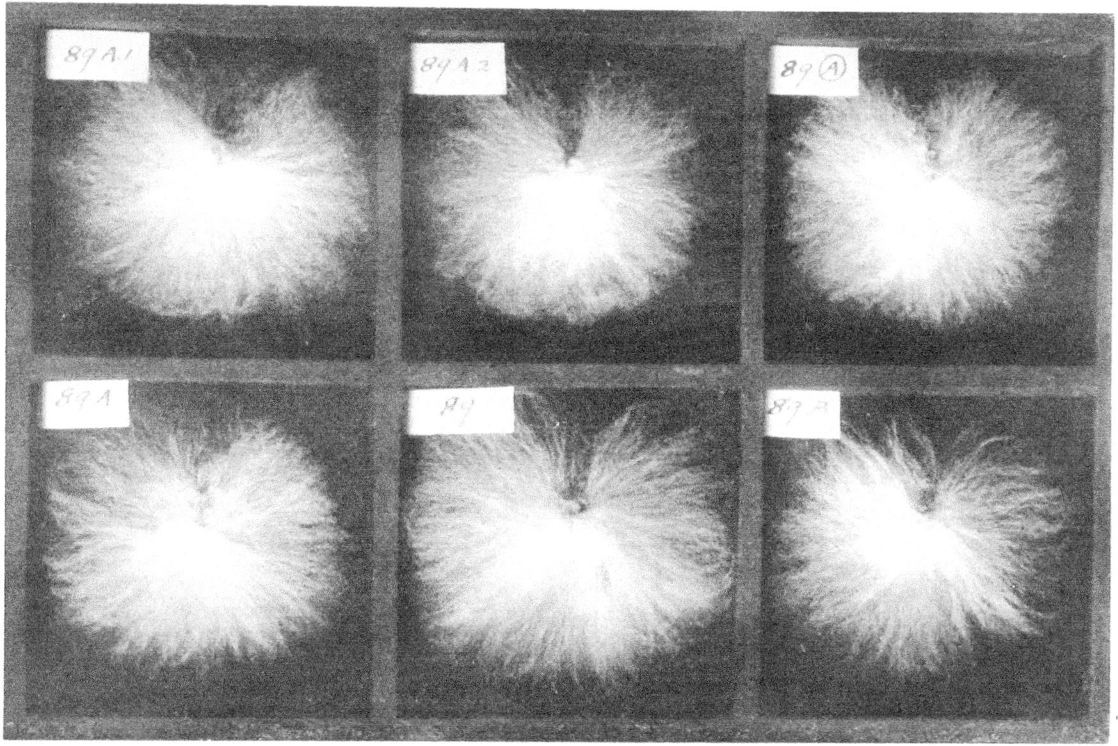

PLATE V.—COMBED SEED-COTTON.

Four-sevenths natural size. The measurements of lint-length used in this book are chiefly made on seeds thus combed. Note the fluctuation from seed to seed within the pure strain, due to external causes.

be crossed with one another. Under field conditions, however, this latter cross is comparatively uncommon; thus, though the Hindi Weed is common in fields of Egyptian cotton, hybrids between the two are comparatively infrequent. This is due to the pollen tube growing faster down the style of its own kind of plant than down a foreign style; consequently, if both self and foreign pollen reach the style of any flower, nearly always the foreign tube will be beaten in the race to the ovules.

Intercrossing under field conditions is usually confined to closely related forms, and from such crossings there arise various recombinations of the factors composing the parents, and consequently new varieties. The more dissimilar the parents are, the greater will the number of recombinations be, and the rarer the chance of such a new form breeding true. Still, if further crossing is excluded, perfectly pure new forms will segregate out from the mixture, and new varieties will result from their multiplication, equal in definiteness to the original elementary species. There is no very definite convention as to the distinction between elementary species and varieties.

The amount of natural crossing which takes place in cotton under field conditions was formerly supposed to be negligible; but the author in 1905 showed that about 5 to 10 per cent. of the cotton-seed in an Egyptian field crop was not self-fertilized, and since then it has been elsewhere shown that most other commercial cottons intercross to about the same extent. The effect of this crossing is gently to mix, and to keep mixed, the pedigree of the plants composing the crop, so that even if a variety consisted of only two elementary species when first introduced, it would soon be complicated.

Origin of Commercial Varieties.

The commercial varieties have usually been derived from single plants, or groups of similar plants, selected from these mixtures. Sometimes they have been bred from an artificial cross, but the difference is slight, unless one of the parents was an entirely fresh introduction to the country of growth, which has seldom been the case. Often they have not been bred down to the pure form before being placed on the market, though externally no marked differences were obvious. In other cases they have been introduced from the beginning in a hopeless state of mixture, such as the Assili cotton of Egypt (Fig. 16, Targets 4 and 8), whereof half the flowers were golden-yellow, half light yellow, and the length and outturn of the lint showed nearly 50 per cent. of rogues in the second year of its introduction. There are indications that some varieties, other things being equal, are more susceptible to crossing than their neighbours, so that the rate of deterioration of a variety varies; but it should now be obvious to the reader that even a very small percentage of impurity in a new variety must ultimately leaven the whole lump. Even if the variety is introduced in an absolutely pure condition, it is bound to deteriorate in the end, owing to the admixture of neighbouring varieties, such admixture being brought about by imperfect cleaning of ginneries—and perfection is commercially impossible—by resowing with other seed, by self-sown seedlings springing up in the field, or by the shooting of rattoon stumps from an old crop, as well as by bees carrying foreign pollen from a distance, and so making bastards with unknown pollen parents.

Deterioration of Varieties.

There is nothing magical or unpreventable about the deterioration of cotton varieties, and every case known can be explained in terms of crossing, seed mixture, and natural selection.

This discussion may appear to have wandered a long way from the cotton fibre, but it is necessary that the principles involved should be understood, because the purification of cotton varieties, and their maintenance in a pure state, is almost the only big advance in the technique of cotton-supply which is economically practicable at the present day. Research is showing more and more clearly that uniformity is the chief thing needed by the trade, and lacking in the field; and while it is quite possible to produce useless rubbish from a perfectly pure strain of ideal properties, perfect cotton cannot be produced from an impure strain. Moreover, cotton is grown by, and dependent upon, cheap indigenous labour. Such labour is not easily reformed to Western requirements, simply because such reformation would not pay the labourer, and the provision of pure seed in place of impure does not interfere with any of his traditional or casual methods.

Seed Renewal.

The cultivation of pure strains of cotton, whether they be selected elementary species, purified existing varieties, or new strains synthesized by crossing, is only practicable with a system of seed renewal. The causes leading to deterioration cannot be avoided in the field, but they can be avoided in the laboratory. Strains can therefore be kept pure on a small scale, and a fresh stock run up into bulk (Pl. VI.) each year to replace the contaminated descendants of previous years. The system was introduced into Egypt by the writer, and at the time of writing is in abeyance through misunderstanding of the complexities involved in the isolation and testing of new strains; but it will inevitably become general before many years have elapsed in all countries supplying fine cottons to the trade. Whether it will ever be worth while to apply it to coarser cottons remains to be seen, but though uniformity is less important in inferior cottons, the possession of it still increases their value, and the question of yield is also worth consideration.

Yield and Breed.

In such a country as Egypt the limiting factor of the yield is the conditions under which the plants are grown, and it is a little doubtful whether any change in the type of plant cultivated could increase the yield appreciably, though the author is inclined still to think that this could be done. In India, on the other hand, and in most countries dependent on rainfall, the conditions of cultivation cannot be so perfectly controlled, and the problem is rather one of how to produce and maintain varieties which will give the best results under the existing average conditions. The researches of Mr. Leake on the breeding of branching habits and of ginning out-turn may be cited as an example of the beginning of such work, and the logical outcome of any such developments must also be the supply of pure seed and provisions for its renewal.

In concluding this chapter, it may be useful to point out a few of the revised ideas which the isolation and study of pure strains have introduced into the cotton trade. Perhaps it would be more correct to say "will introduce," for the disposition to regard cotton-plants and varieties as capable of reasonable behaviour has yet to be manifested.

Modern Ideas.

In the first place, the greater part of this book is occupied by results which could not have been obtained, in their present form, on a mixed commercial variety. One experiment describes the steady oscillation in length of lint between 34 and 31 millimetres (seed-combed length); the most uniform commercial variety in Egypt contains plants which range from 25 to 33 millimetres on the same day (p. 134, Target 11); so that such slight changes as an eighth of an inch (3·1 mm.) are almost entirely obscured, unless an intolerable amount of additional labour is expended in accumulating data so as to smooth out these constitutional differences between plant and plant.

Precision.

Again, cotton has for generations been held up to censure and admiration alternately as the most "variable" of organisms, capable of being moulded into any form, and equally incapable of retaining it. This opinion has had to go by the board, with the progressive analysis of the phenomena into such constituents as those outlined in this book, and the plant is now known to be no more variable than any other, and equally controllable. Two pure strains of cotton grown by the writer may be cited: one

Variability of Cotton.

had remained unchanged for nine years, the other for seven; one would not produce half a crop in the north of Egypt, the other would not produce half a crop in the neighbourhood of Cairo; one consistently contained about 30 per cent. more salt in its cell sap than the other when the roots were occupying the same soil. Such simple differences as $\frac{1}{4}$ inch in lint length, 1 per cent. in ginning out-turn, and 2 or 3 degrees in the angle of the lobing of the leaf; colour of leaves and flower, shape and dimensions of the boll, fuzziness of the seed, and so forth, while all capable of being distorted from the normal, were, it need hardly be added, all showing exactly the same under the same conditions in 1913 as they had been in 1907. Left exposed to crossing and mixture for a single season the strains showed 10 per cent. of impurity in the following year.

Interpretation of Popular Expressions.

Such strains as those mentioned, in conjunction with other experimental evidence, throw light on such customs as "change of seed," "selection for yield," etc. From a sowing mixture of these two strains in equal parts we harvested 75 per cent. of one or the other, according as to whether the mixture was grown in the south or the north. In the following year the seed from this mixed sowing produced a percentage of hybrid plants; selection of the highest yielders in the population included nearly all these hybrids, and scarcely any of the parental stocks. Thus in the third year, had the seed of these plants been sown, both the pure parents would have been lost entirely, and a most intricate jumble of all sorts of cotton would have rewarded this blindfold selection.

CHAPTER II

THE DEVELOPMENT OF THE PLANT

REFERENCE to many standard works on the subject will provide the reader with full descriptions of the various kinds of cotton-plant; but these descriptions, though accurate and invaluable, do not convey the impression of "livingness." In this chapter we shall therefore attempt to describe the main features of the plant in a somewhat different fashion, presenting rather a kinematograph than a simple photograph of the cotton-producing machine. From this we may proceed to details in the ultimate stages of lint formation.

It may be well to remind the reader that, although this account of the life of the plant* is very largely generalized, and of universal applicability in principles, the mental picture before the author is mainly—though by no means exclusively—one of Egyptian plants growing under Egyptian conditions. Thus, water is associated in the writer's mind with a controlled irrigation rather than with rain, extreme heat with temperatures from

* Those readers who, having no knowledge of Botany, are interested in these aspects of Cotton, are advised to read "The Life of the Plant," by Professor C. A. Timiriazeff, in the English translation. (London, 1912.)

100° F.* to 115° F., and cold with such temperatures as 40° F. The tracking of cause to effect is necessarily easier in Egypt than in some countries—as regards the cotton-plant—but the same interpretations are being found to hold good elsewhere, though with greater complication; and it might be well to notice here that even Egypt is far from possessing a monotonous climate, excepting during July, August, and September. A difference of 20° F. between the maximum temperatures on successive days is not at all uncommon in the other parts of the season. Further, this limitation is less objectionable than it might be, because the researches presented in this book are of more interest to the fine-spinning trade than to the ordinary trade, and Egypt is the largest producer of such fine cottons.

The characteristic feature of all cotton-plants is the bell-shaped flower (white, yellow, yellow with red spots, or entirely red), with a brush of golden or buff stamens borne in the centre on a hollow cylinder, through which the style extends to the exterior from the ovary, and surrounded externally by three leafy bracts (Pl. III.). The latter feature distinguishes cotton from such plants as *Lavatera*, which is frequently confused with it, and when the capsule has opened into two to six loculi, exposing the white or brownish cotton, the plant is unmistakable. The main differences between the principal groups of cottons have already been indicated.

The stages of the life-history before sowing have very little immediate interest for us, and we may take up the

* 100° F. to 115° F. = 38° C. to 46° C.; 40° F. = $4\frac{1}{2}$° C. For conversion of future statements of temperature, note that $(C.^\circ \times 1\frac{4}{5}) + 32^\circ = F.^\circ$

story at the stage when the seedling is well established, with several leaves, and a field of them is beginning to bear a rough resemblance to a crop of potatoes.

Stunting.

If the seedlings before this stage have been stunted through any cause, such as imprisonment under clods or in stiff soil, through the bite of a prowling caterpillar or the rotting action of fungi, by heat or by water shortage, they will be smaller than their neighbours, take longer to open their first flower, and will therefore—other things being equal—yield less cotton, simply because they have less time in which to open and ripen their flowers.

Sowing-Time.

Similarly, seed which is sown too late will yield a smaller crop for the same reason, though the plants may grow better than their early-sown neighbours, stage for stage. It does not follow, however, that sowing very early is of any advantage; it may be pernicious, and in Egypt there is a Critical Sowing-Date in each district, which varies only a few days from year to year. Sowings made moderately early before that date all do equally well, while sowings made afterwards are proportionately later in flowering (Fig. 2). It is obviously best to sow on the critical date itself, since the chances of a cold spell are less as the summer approaches.

That this should be the case is obvious on a little reflection. Cotton is usually sown when the weather is getting warmer (when the reverse, as in the Sudan, the same arguments apply), and there are two ends to the plant, of which the root is equally important with the

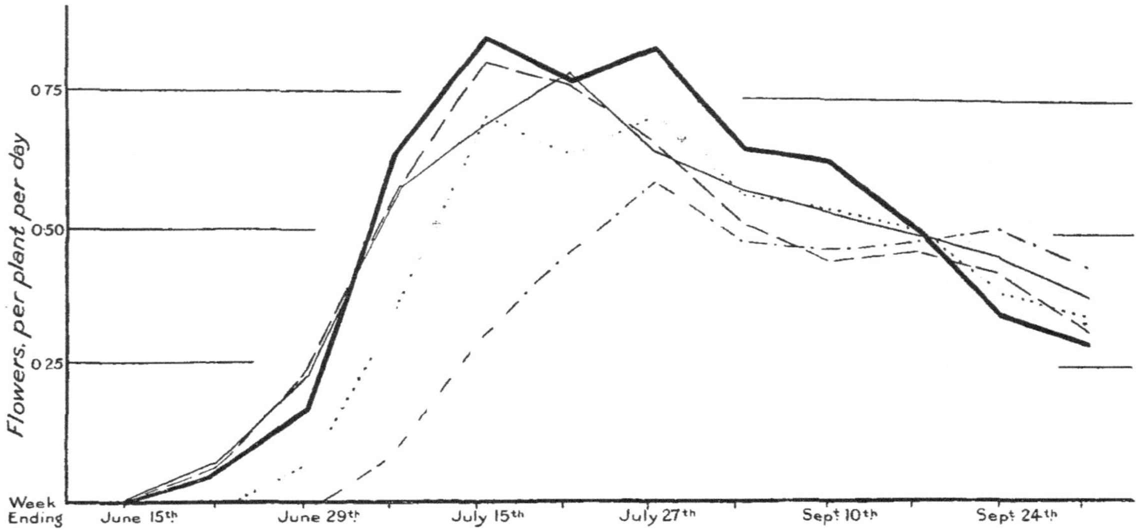

FIG. 2.—EFFECT OF SOWING TIME ON FLOWERING.

The five curves show the average number of flowers opening daily in each week of the flowering season. Each curve represents the average behaviour of 1,000 plants arranged in five scattered plots (see Pl. XI.). Variety, *Domains Afifi*; site, Giza; year, 1913.

— — = Sown one month *before* the usual date. ——— = Sown a fortnight *before* the usual date. ▬▬▬ = Sown *at* the usual date (March 15).

These three all come to maturity at the same time, but since a certain number of plants have been stunted by accidental circumstances, due to cool weather in the earlier sowings, the sowing at the usual date is actually the best.

...... = Sown a fortnight *later* than the usual date. — · — = Sown a month *later* than the usual date.

These develop more quickly, but cannot make up all the time lost, and, being late in flowering, ultimately give a smaller crop.

shoot. The rate of growth of the root is usually controlled by the temperature of the soil, provided that the soil is sufficiently moist, and that the stem is sending enough food to it from the leaves. In the early spring the soil is too cold for the root to function freely, and the stem suffers in consequence, though the air may be warm enough. Soil temperatures at a few inches below the surface scarcely vary their annual change from one year to the next, so that the date of sowing depends on deep soil temperature in the first instance, which is the same on the same day in the same field each year.

Further, to sow excessively late will result in failure of the seedlings through overheating. This phenomenon recurs at later stages of the plant's history, and merits some attention. Seed germinated in incubators at various temperatures will provide simple illustrations. At 15° C. the germination is slow, while at higher temperatures the rate of growth increases; between 20° C. and 30° C. the rate of growth is doubled, and if an incubator is adjusted to 36½° C., it is possible with Egyptian cotton to prepare a report on the germination capacity of a sample within twenty-four hours from receiving it. If, however, a sample thus incubated is left for two days, it will be surpassed by those samples kept at lower temperatures, and in three days will be almost irreparably injured. This injury is due to accumulation of poisonous excreta in the tissues, these being thrown off in the chemical processes of growth more rapidly at high temperatures than the rate at which the plant can dispose of them.

Poisoning by Heat.

Similar poisoning from overheating takes place when seed is sown too late into the summer. Otherwise the root rarely suffers from this cause, being buried in the cooler soil ; but the stem is often affected on hot days, when the temperture exceeds 37° C. (Fig. 4). It must be remembered that the actual temperature of the stem tissues is the important thing in this respect, and not the shade air temperature, which may be widely different, the plant having a lower temperature than the air, through evaporation from the leaves, when there is ample water round the roots, and a higher temperature if the roots are short of water. The presence or absence of wind, clouds, newly watered soil, etc., also affect the temperature of the air itself in the cotton-fields, and may make it quite different from the temperature recorded in a meteorological screen close by.

The Temperature of the Plant.

We have seen, then, that a rise of tissue temperature up to about 33° C. continually accelerates growth, but that prolonged exposure to higher temperatures than this is prejudicial. This all assumes that light, water-supply, etc., are sufficient. If any one of these is deficient, the growth will accelerate with rise of temperature until it reaches a rate at which it is using all the food or all the water available, and beyond this rate it plainly cannot pass, even though the temperature continues to rise; in other words, food or water becomes the "Limiting Factor" instead of temperature (Fig. 3).

Limiting Factors.

The relations of the seedling to water bring out one or two points of interest, which bear upon our previous conjectures as to the utility of the lint to the plant for absorbing and retaining water instead of acting as seed-wings. Cotton-seed germinates best when lying in water, half covered. If laid on wet blotting-paper in the conventional way for making tests of germination with many other seeds, very

Water and Germination.

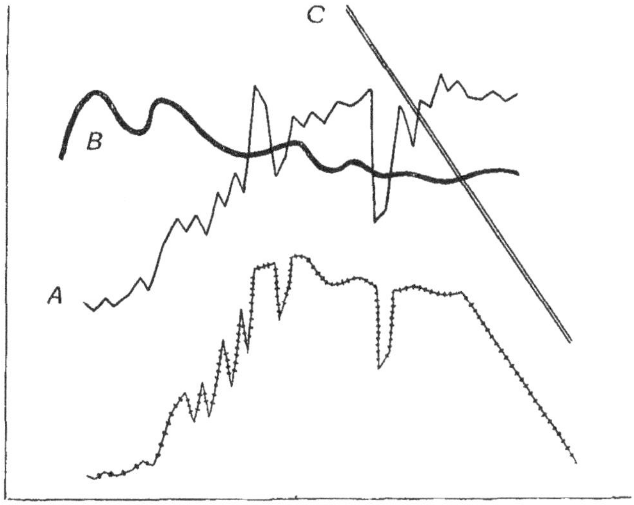

FIG. 3.—THE ACTION OF LIMITING FACTORS. (VERY DIAGRAMMATIC.)

Three varying factors are indicated above (*A*, *B*, and *C*) as curves. Below is represented the behaviour of a plant under their influence, as first one and then another becomes "limiting."

If *A* represented the temperature, *B* the water-content, of the soil, and *C* the size of the living root-system (as affected by the water-table), this diagram would roughly portray the life-history of an Egyptian cotton-field.

few of the seeds germinate, whatever the temperature, the seed-coat being unable to absorb enough water.*

* The cuticle of the seed-coat is slightly waxy, probably from the same wax which has long been known to occur on the lint. Until this wax has been broken by the initial swelling of the seed, the absorption of water is difficult.

This cause of germination failure is common in some countries, though not in Egypt; but even there it may be found. Seed sown in very lumpy soil will not be thoroughly wetted for a sufficient time, and seed sown too early in the spring may fail from the same cause. The latter is a rather curious point. It would at first sight appear that cold was the limiting factor, but neighbouring seeds will come up freely, and the failures themselves will germinate on a second watering. The cause of failure is simply that at the lower temperatures the absorption of water by the seed-coat is too slow, from purely physical causes, unless the seed is thoroughly and continuously in contact with soil-water.

Water and the Plant.

The reader may have noticed that, in speaking of the cooling effect of evaporation upon the temperature of the stem tissues, we phrased the necessary condition as "ample water around the roots," and not merely as ample soil-water. The reason for this is that the root may dry the soil-particles which lie near it. The author first showed in Egypt that the cotton-plant stood in a rather extreme attitude towards its water-supply, which might almost be described as one of greed and improvidence. These results have since been confirmed by Professor Lloyd for the United States with the Upland crop, so that there is some justification for believing that they are general. The phenomena are most marked in the later stages of the plant's growth, and do not begin to exert much limiting control until after flowering has begun, provided that the seedlings receive a reasonable water-supply. The reason for this is simply the increasing area of leaf-surface, causing greater loss of water. If, however, seedlings or young plants are allowed to become very short of water, they may be limited thereby, with subsequent effects on the crop, to which we shall later advert.

The basis of these marked peculiarities of the plant in regard to water is a structural one—namely, the presence of a rather exceptionally large number of "stomata" on both sides of the leaf. These breathing-pores, which also act as lip-valves for regulating the loss of water-vapour, number about 300 to the square millimetre on the lower leaf-surface (or 200,000 to the square inch), and 100 on the upper surface. In cottons of the Peruvian type the surface is practically hairless, but in the Upland type the leaf-surface is commonly hairy, and since the hairy tangle prevents rapid motion of air past the exterior aperture of the stomata, evaporation is diminished. Upland cotton can in consequence endure dry weather with less injury than Egyptians—a fact which will have some influence on the development of cotton-growing in many new areas. Some idea of the number of these apertures may be gathered from the fact that a seedling without any other leaves than the pair of original seed-leaves is pierced with about a million stomata. The evaporation of water takes place almost entirely through these apertures in the older leaves, though in very young tissues, which have not developed the impermeable skin of cuticle properly, some water escapes directly from the skin of the leaf. The stomata open and close in reaction to the condition of the plant. If water is deficient, they close before the plant is noticeably wilting, and so restrict further loss.

Under the climatic conditions in which cotton finds its most suitable temperatures, the mere evaporation of water from a water-surface is usually very high during the day, and the total "water-surface" of all the leaves in a field of cotton is enormous. Some idea of its magnitude may be gathered from the fact that the author and Mr. Hughes, working quite independently, both showed that normal fields of Egyptian cotton in a normal year (1912) evaporated amounts of water which increased steadily up to *fifty tons* of water per acre per day. This figure is almost incredible, being twice as high as the amount of water actually supplied to the land in irrigation—the duty of water in Egypt being twenty-four tons—but the deficiency is made up by water withdrawn from the water-table. When expressed in terms of single plants, it is even more incredible—*i.e.*, about three pints per plant per day.

Water and the Crop.

There can, however, be no shadow of doubt as to the truth of the figures, which in both cases were obtained by directly measuring the changes of water-content in the soil of a field. They throw into vivid relief the severe nature of the "water-strain" which the plant has to undergo each day of its life.

One of the first evidences of this strain was the writer's discovery that from quite the beginning of its career the cotton seedling did not grow in sunshine, all the water being used in keeping the plant cool as long as possible; and when the soil supply began to give out, at some time during the afternoon, the stomata closed.

Sunshine Effect.

This closure of the stomata brings about other effects which may be roughly described as starvation, since carbon dioxide gas can no longer pass into the leaf, where it would be built up into sugar for the food of the plant. The result is that the plant, as a result of its water-greed, leads a rather miserable existence every afternoon in hot sunny weather.

One Day's Events.

The story of its average day is roughly thus: Stomata open with the sunrise, and open wider as the light gets stronger, the formation of sugars beginning and increasing as the day gets warmer, till carbon dioxide may be taken up as fast as a free surface of caustic potash could take it, and the weight of the leaves increases rapidly. Growth has meanwhile slowed down or stopped entirely, according to the wetness of the soil and the humidity of the air. Presently, at some hour after 9 a.m., depending on the same conditions, the plant has dried up the soil in the immediate vicinity of its roots by taking away more water than can be replaced by capillary movements among the soil particles, and the stomata begin to close, while growth is entirely stopped. This closure of the stomata checks the formation of sugars, by cutting off the supply of carbon dioxide, and the transport of sugars, etc., into the body of the plant being no longer compensated by the formation of new supplies, the dry-weight of the leaf decreases. The plant remains thus during the afternoon until sunset, when—the water-loss ceasing—growth is resumed, often quite suddenly. During the night the rate of growth is controlled by the night temperature

(Fig. 4), until the food formed during the day is exhausted, which may happen before the morning if the previous day's experiences have been very severe, though this does not appear to be often the case in good cultivation. The total growth in the twenty-four hours thus depends mainly upon the temperature at night, with modifications. The chief of these modifications are growth in the morning after sunrise, due to milder weather ; and overheating during the day, which poisons the cells and slows growth during the following night (Fig. 4).

Suitable Soil.

Lastly, of the three main controlling factors of growth, we have, in addition to temperature and water, the soil itself. In so far as texture is concerned, it is easy to read from the foregoing discussion the reason which makes loam the most suitable soil for cotton—namely, free movement of water to the root system, superadded to ample retention of water by the soil particles. There is a second reason, which is that cotton roots appear to be intolerant of any deficiency of air for their respiration (probably for reasons connected with those discussed) such as may easily happen in a clayey soil.

Spacing of the Plants.

Besides the texture of a soil, its depth and its composition have to be considered. These two go partly by inverse proportion. A large plant will flourish in a small pot of rich soil or in a large pot of poor soil; but the daily water-strain on cotton-plants makes a large volume of soil essential for the best results, unless unceasing watering can be given in driblets, which is not practicable in the field. A large volume of soil might result from placing the plants widely apart, but this, beyond a certain lower limit, would reduce the yield per area, since the increase in size of the plants

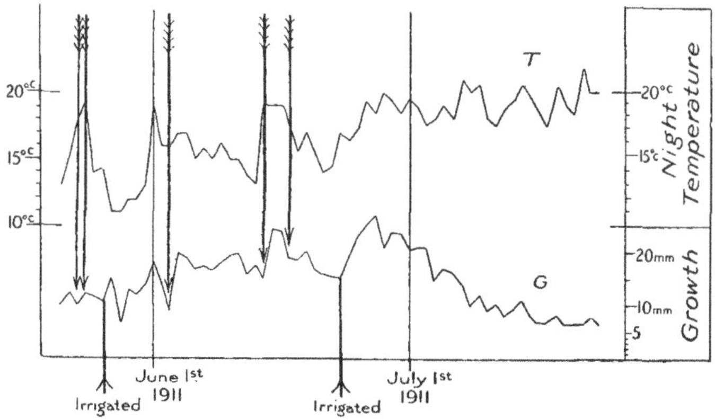

FIG. 4.—DAILY GROWTH OF THE MAIN STEM.
Actual experimental data.

Variety, Afifi; site, Giza; ordinary field crop; year, 1911. *T*, upper curve, represents nightly *minimum temperatures*; *G*, lower curve, represents the amount of elongation of the main stem in each successive twenty-four hours.

Until the end of June the two curves are similar—*i.e.*, growth was limited by night temperatures, except upon days which are marked in the diagram by eight feathered arrows.

These arrows denote days when the *maximum* temperature in the previous afternoon had been too high (over 36° C.), so inducing the formation of a "heat-poison" in the plant, and thus reducing growth during the following night to less than the amount expected (indicated by dotted lines).

At the end of June the effects of soil-water were just beginning to be felt as the plants grew larger and took more water from the soil. Previously there had been more water than the plants could remove.

Further, this shortage of water induced "self-poisoning" in the terminal bud of the main stem, and its growth ceased ultimately to respond even to irrigation. The terminal buds of the branches are similarly affected later.

would not compensate for the smaller number of plants on the area (Fig. 6), and, moreover, surface soil dries up more quickly than deep soil, so that it would be harder to keep a uniform moisture-content. While discussing this soil question, it may be worth while to note that the effect of spacing upon yield is worth considerable attention by new cotton-growing countries. The author showed in Egypt—somewhat to the disappointment of reformers—that the native cultivator is exactly right in the spacing he adopts (Fig. 5). If more plants were crowded on the area, they would, roughly speaking, interfere with one another's roots before they came into

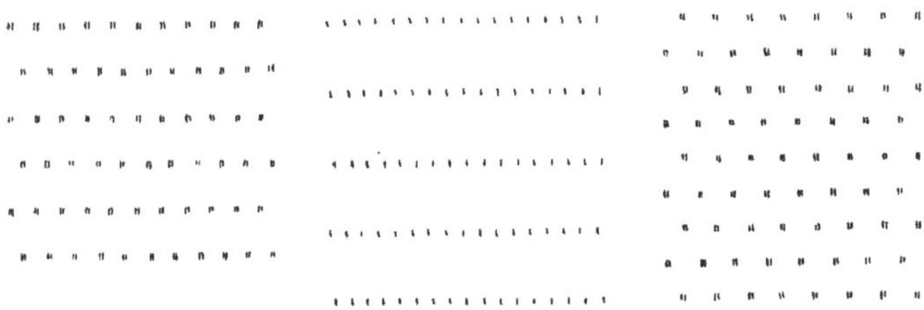

FIG. 5.—ARRANGEMENTS OF THE SPACING IN FIELD CROP.

Plans showing on a scale of $\frac{1}{192}$ the typical conventional arrangements of the plants in the field in Egypt (left), and the U.S.A. (centre), together with (right) an arrangement which would be ideal if it were not impossible to hoe, irrigate, and pick, such closely sown ridges.

The Egyptian arrangement is only possible because the crop is entirely cultivated by hand. In the U.S.A. the rows must be set wider, to permit of the use of horse-hoes (*cf.* Pl. XII.).

flower, and would suffer more under the water-strain of the early autumn, and from the rise of the water-table. If fewer plants were employed, the gain in yield per plant from the diminished effect of these same causes could not make up for the decreased number of plants actually at work. At the spacing which the Fellah employs in each district, the effects of root-interference are delayed till the first few flowers have formed, and the gain thus obtained in the early picking (Fig. 6) through having more plants is nicely counterpoised by the loss in the later stages through having too many roots in the soil.

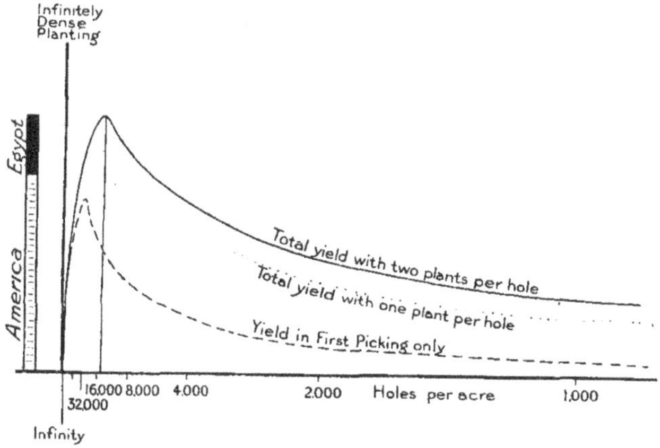

FIG. 6.—EFFECT OF SPACING ON YIELD.

These curves are generalized from experimental data obtained at Giza in 1912 (see Phil. Trans. Roy. Soc., 1915, for the full account).

The tall vertical line on the left marks the extreme limit, or "infinite density," when the seed would be so thickly planted as to give no yield at all.

The maximum total yield is obtained at the conventional Egyptian spacing of about 14,000 holes to the acre, with two plants in each hole.

The employment of one plant in each hole lessens the total yield by about 10 per cent. (dotted line) in any spacing.

At wider spacings than 1,000 plants per acre the yield is directly proportional to the number of plants.

The first few flowers to open are directly proportional to the number of plants on the area, unless the spacing is extremely dense; thus, the first picking (dash line) is at its largest with a closer spacing than that which gives the maximum total yield.

The curves here drawn apply to fairly symmetrical spacings, such as the ideal and Egyptian ones in Fig. 5, opposite. Any departure from symmetry, as in the U.S.A. arrangement, diminishes the yield, other things being equal. The difference between the total yields with Egyptian and U.S.A. arrangements, but with the same number of holes per acre, is about as 5 : 4. This is shown graphically to the left of the diagram, where the black top of the column represents the advantage of the hand-cultivated Egyptian system over the American one.

It being agreed that the larger the volume of soil available, the better it is for the plant, and lateral extension being ruled out by considerations of yield, it follows that the soil must be deep. In respect of this our ideas have undergone great changes since 1908, in which year the author first traced a cotton root to a depth of two metres below the surface. This trifling episode led us to reconsider many matters. Thus, in the event of surveying a new district for cotton-growing, we are no longer content to examine the top foot of soil, but we require to know something about the state of affairs down to some 6, or even 10, feet below the surface. Again, that a rise of the water-table which did not reach within a metre of the surface should be prejudicial to the crop was no longer absurd, since this would drown out half the root system of an adult plant, and, as we now know, the portion which is doing most of the work in the autumn. It had formerly been thought that cotton was a comparatively shallow-rooted plant, on account of the slenderness of the tap-root, which diminishes very quickly in the first foot of descent, and looks as if it were nearly at an end. This slenderness does not prevent it from carrying large amounts of water, as we have seen, the main water-conducting cell-elements being only two or three in number, and consequently of wide bore. Through these two or three tubes, not exceeding $\frac{1}{2}$ millimetre each in diameter ($\frac{1}{50}$ inch), there rushes a stream of about a litre of water during some ten hours. From the tap-root arise laterals, which at first form a gossamer web around the seedling tap-root, but which very largely die down, only a few surviving, to be supplemented by new ones lower and lower down. The survivors run out

The Root.

horizontally almost as far as the tap-root descends vertically, though not quite so far if other things are equal, and are important contributors to the plant's sustenance until the soil layer which they occupy is too crowded with roots from adjacent plants to be of much use.

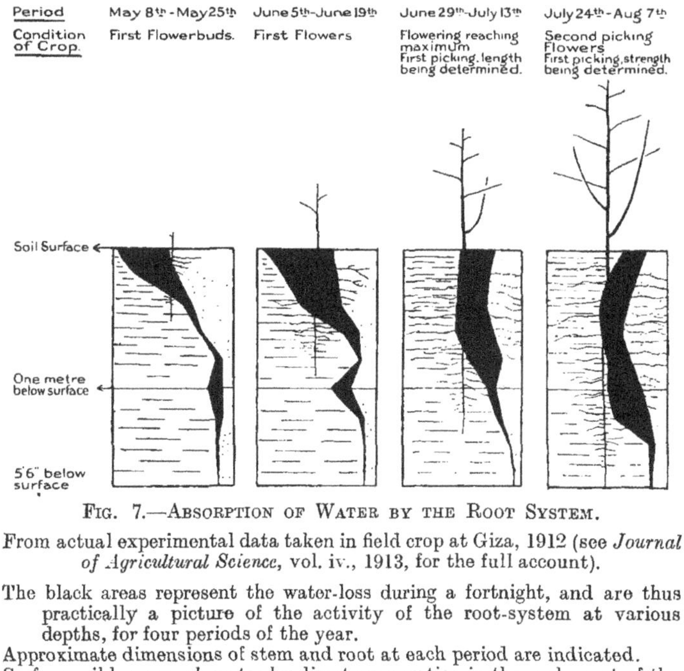

FIG. 7.—ABSORPTION OF WATER BY THE ROOT SYSTEM.

From actual experimental data taken in field crop at Giza, 1912 (see *Journal of Agricultural Science*, vol. iv., 1913, for the full account).

The black areas represent the water-loss during a fortnight, and are thus practically a picture of the activity of the root-system at various depths, for four periods of the year.

Approximate dimensions of stem and root at each period are indicated.

Surface soil loses much water by direct evaporation in the early part of the season. Later on the plants shade the surface, and most water is derived from three to four feet below the surface at the time when the first picking is ripening.

The zone of most importance in the root system thus shifts steadily downwards as the season goes on, until, at the time when the first picking is ripening in Egypt, the crop is actually taking more water from a depth of four feet than from any other part, so that a boring-tool may be pushed through wet mud on the surface and broken by attempting to struggle through the hard dry soil at that depth (Fig. 17). Many other side-issues show the enormous possible extent of the cotton root system. Thus, a plant which is allotted 3 square metres of surface will produce more flowers and more cotton than a plant which receives 2 square metres allowance. Since the root of the latter has about five tons of soil to itself, it might be considered to receive ample accommodation; but the plant can still make use of more. The last fact brings up in vivid relief the artificial conditions of field-crop cultivation, where the plants are crowded together (Fig. 5, Pl. XII.) to such a density as will produce the maximum yield, and the consequent limitations in the size of the root system have to be met as far as is possible by skill in cultivation. The author is sometimes inclined to think that in irrigated countries with high-priced cottons the tendency will be towards even closer crowding of the plants, as skill in cultivation becomes greater through better understanding of their necessities. Certainly the heavy first picking which results from closer spacing should be a valuable asset (if skilfully ripened off) in countries troubled by boll-worms, which attack the later pickings.

Deep Draught of the Root.

The last component of the soil question is the chemical composition, and here Egyptian experience is not of much use, since manurial composition is rarely the principal limiting factor of growth there, though there is some indication that it acts as such for short periods at certain stages of growth in the summer. On the other hand, the presence of salts in the soil is often a serious factor in Egyptian cotton-growing, whether the salt is common sodium chloride or the more objectionable carbonate, or "black alkali." The action of salts is relatively simple; an excess of them prevents water-absorption by the root through simple physical control, while less amounts may in some cases act poisonously. Under field conditions one or two interesting points arise. Since salt is concentrated at the surface of the soil by evaporation, and washed down again by surface watering, the surface soil may be so salt as to hinder the germination of seed; but any odd seedling which may manage to work its tap-root down into the sweeter soil below will flourish. Thus, a perfectly healthy plant may be growing in the middle of a salted patch of land. Another curious feature of cotton with regard to common salt is that it takes up quite large amounts from relatively sweet soil, the concentration attained in the cell sap being different with different kinds of cotton, and amounting in the most "salty" kinds to nearly as much as in typical salt-marsh weeds which are washed by spray from the sea.

Alkali Soils.

In most countries other than Egypt the manurial composition of the soil is an important factor, but one which has never been properly investigated, with the result that most conflicting opinions are held. In saying that no proper investigations have been made, the author would not wish to discredit the large amount of experimental work which has been carried out in this

Manures.

direction; but the fact remains that it is almost impossible to analyze the results of such experiments (*cf.* Fig. 19). So many circumstances may act on the plant, and the water relations are so important, that to disentangle the manurial effects from the accidents, over a cropping period of two months, is almost impossible. Solution of these manurial problems can only be obtained—except in the simplest cases—by keeping continuous records of the growth, flowering, and fruiting, of the plant day by day, so that the action of the various factors may be distinguished from one another. This is more particularly the case with regard to the many cases in which it is stated that manuring diminishes the yield through causing too rank growth. The author has never yet met with a case under Egyptian conditions in which growth was ever the cause of reduced yield in itself. Secondarily, it may lead to the exaggerated influence of external causes, if cultivation is not, or cannot be, modified to meet the new dimensions of the plant; but it is much to be hoped that a more thorough analysis may ultimately relegate this view to the limbo whither several venerable fictions have already been despatched, such as the dictum that "the longer the staple of cotton, the lower must be the yield." We will return to some of these points subsequently.

Thus far we have attempted to give some idea of the interplay of circumstances which act upon the living body of the plant, modifying the growth of its various portions, and building up the scaffolding upon which the cotton is borne, as well as the fruit and the cotton itself.

The ground we have covered would require several large volumes for any adequate presentation, even of the established facts, unless such presentation were made in severely technical language; but enough has probably been said to show that there is no longer any need to indulge in vague generalities about vitality, suitable climate, and so on, and that it is possible to assign numerical values to the quality of the cultivator's work. It now remains to examine the scaffolding itself, and its load of fruits, built by the plant under the control of its inherited tendencies as reacted upon by environmental agencies, and itself continually acted upon in the same way.

Plant Scaffolding.

And here it may be excusable to insert a small comment on the difficulty which many persons feel in accepting recent interpretations of this reaction. Their objection may be summarized in such a case as the following: They have grasped the idea of the chemical definiteness of a pure strain, characterized, let us say, by a 30-millimetre lint; but when they find that a 35-millimetre lint is obtained from this strain on cultivation under some particular set of conditions (*cf.* Figs. 14 and 15), they object that it must be nonsense to speak of any definiteness in the inheritance or manifestation of such variable characteristics. A simple parallel will best illustrate the falsity of the objection. Certain strains of the Chinese primrose are known which breed true to white, whatever the conditions under which they are grown; there are also other white strains which remain white when grown in a cool greenhouse, but turn pink

Variable Characters.

in a house kept at 30° C. The inherited peculiarity in both cases is not actually colour, but the absence or presence of the power to react under environmental influence—high temperature in this particular case—so as to produce a colour.

So the plant inherits the capability of reacting in a certain exact and definite way to any set of exactly defined conditions.

The plant-scaffolding of cotton consists of root and

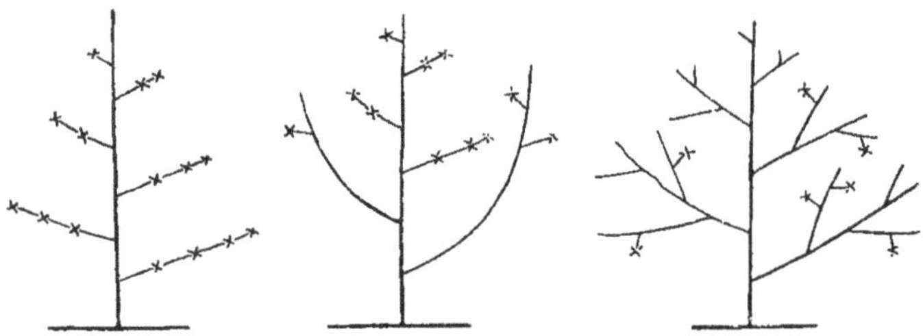

FIG. 8.—TYPES OF BRANCH-SCAFFOLDING. (DIAGRAMMATIC.)
Flowering branches are represented as lines bearing crosses, vegetative branches as plain lines.

Left: Ideal plant, early flowering (*cff.* Pl. VII., X.).
Centre: Usual type of plant (*cf.* Pl. IX, right).
Right: Late flowering. Vegetative branches branch again before flowering branches are formed (*cf.* Pl. II.).

shoot, but very little is yet known about the details of the former, owing to the obvious difficulties which attend investigation. The shoot-scaffolding begins as a main stem, or central axis, from which lateral branches are given off. These branches may be flowering branches (Pl. VII.) or ordinary vegetative branches; which latter may again produce other vegetative branches, or flowering branches, or both, according to the inherited tendencies of the plant (Pl. IX.).

Branches.

PLATE VII.—SIMPLE BRANCHING.

A plant of a very atypical strain isolated from Abbassi. Wide-sown, at Giza, September, 1913. One-fifteenth natural size. Scarcely any vegetative branches produced (*cf.* Plate IX.). The ripened bolls of the first picking show the "fruiting-scaffolding" of the cotton-plant in its simplest form (*cf.* Figs. 8 and 9, pp. 42 and 44).

The flowering branches, on the other hand, bear a limited number of flowers, and do not normally yield any further branches.

The development of these various branches is mainly acropetal and centrifugal, or, in other words, a branch high up the main stem is younger than a lower branch, and the flowers at the outer end of a flowering branch are younger than those nearer the main stem (Fig. 9). Exceptions to this occur in the relative times of development of the two kinds of branches, but in general it is safe to assume that the opening of flowers on the plant body pursues a kind of spiral course, beginning at the innermost flower of the lowest flowering branch, and ending at the upper and outermost flower of the youngest branch.

The relations between the two kinds of branches are of the utmost commercial importance. If any particular

Cotton producing Machines.

kind of cotton develops nothing but flowering branches on the main stem, it will obviously be an economical machine for cotton production (Pl. VII., X.) if its leaf area is sufficient to feed all the flowers which it sets, until they open as ripe bolls. If, on the other hand, another kind of cotton does not produce flowering branches on the main stem at all, but waits to bear them on the vegetative lateral branches, it will be slow in coming to maturity, and unsuitable for any district which can provide only a short growing season. This problem has been faced by Mr. Leake in India. Into the further details of branching it is scarcely necessary to enter, but it might be noticed that excessive vegetative branching is undesirable, in that it makes a field of cotton into a damp jungle, in which various secondary troubles may arise; while a plant with a large leaf area is more likely to have its physiological condition seriously upset by water-strain in the event of a spell of dry weather.

The length to which the main stem and the various

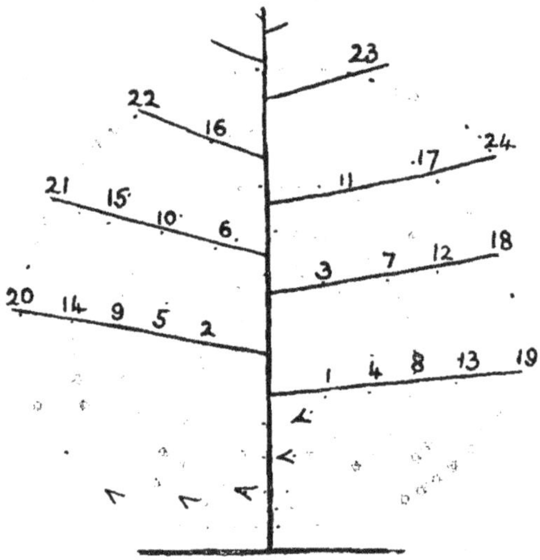

FIG. 9.—SUCCESSION OF FLOWERS. (VERY DIAGRAMMATIC.)
Representation of the way in which the flowers—and consequently the fruits—open successively on the scaffolding of branches.

branches extend is obviously dependent on the various causes affecting growth-rate, as already sketched by us.

Senescence.

Over and above this, however, there comes sooner or later in the career of each branch a period of "Senescence," due to internal causes akin to the heat-poisoning which we have mentioned, and which we may term "self-poisoning." This phenomenon shows itself in a general slowing of the growth-rate, and in diminished reaction to external changes, such as temperature; it is nearly akin to muscular fatigue. Usually the oldest parts of the plant—or, rather, the terminal buds of each part—are the first to show it, and the main stem first of all (Figs. 4 and 10). This cessation of growth of the main stem has an effect on the flowering later on, for, since no more flowering branches can be formed in an upward direction, flowering must stop when all the existing flowering branches have opened all their flowers, unless lateral vegetative branches exist to take up the work.

Like all other features of the plant, although this senescence may be induced by ill-treatment at any time, it is a specific reaction which varies with different kinds and strains of cotton-plants. Under the same conditions of cultivation some kinds show it at a very early age in the terminal bud of the main stem, and are commercial failures thereby; others show it so late that the flowers they have formed are as many as can be ripened off before the winter comes; and others may not show it at all, but continue year after year to respond directly to the limiting factors of the environment, and be known as "tree cottons" (Pl. II.)

Moreover, senescence is not irremediable with cotton-plants as it is with mankind at present. A senescent bud may recover after a prolonged rest, and probably we shall be able to obtain such recovery almost instantly when the nature of the poisons is known. The result of field recovery is a second crop of flowers or even fruits in a single season. Before a regular water-supply was secured for the Egyptian crop during the summer, senescence following water shortage, and rejuvenation when the flood came down, were probably much more normal phenomena than they are at the present day. We shall meet with the same senescence in the fibre itself.

Rejuvenation.

Until this senescence is shown individually by the various branches of the same plant, all branches grow at nearly the same rate. Thus, measurements of the daily growth-rate of the main stem in the early part of the season show (Figs. 4 and 10) the variations in the rate at which the scaffolding of flowering branches is being laid down. It therefore shows also the daily variation in the rate at which flower-buds are being formed on this scaffolding, and consequently—since each bud takes a fairly uniform time to develop into a flower—this pre-senescence growth-record of the main stem anticipates the daily variations in the rate of flowering. In Egypt this leads to remarkable possibilities of forecasting flowering from growth, and consequently of forecasting the crop, since the variations in the rate of flowering in the early part of the season are the same all over Egypt from day to day, some days being good and others bad, mainly as the result of rapid or slow growth during warm or cold nights, nearly a month previously.. How far this will apply to other countries remains to be seen.

Crop Forecasts.

The point of especial interest up to which we have been

leading, through this conception of a flowering scaffolding, is that the ultimate yield of the plant has been partly determined some two and a half months before it appears as ripe cotton bolls.

Various investigations have shown that, as most

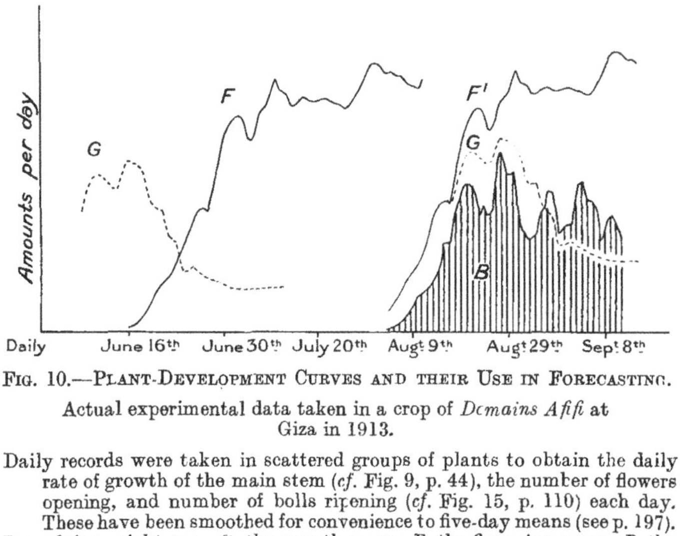

FIG. 10.—PLANT-DEVELOPMENT CURVES AND THEIR USE IN FORECASTING.

Actual experimental data taken in a crop of *Domains Afifi* at Giza in 1913.

Daily records were taken in scattered groups of plants to obtain the daily rate of growth of the main stem (*cf.* Fig. 9, p. 44), the number of flowers opening, and number of bolls ripening (*cf.* Fig. 15, p. 110) each day. These have been smoothed for convenience to five-day means (see p. 197).

From left to right are—*G*, the growth-curve; *F*, the flowering curve; *B*, the bolling curve, the area enclosed by this last having been shaded with vertical lines to emphasize its importance as representing the actual final yield.

If *G* is moved forward for seventy-four days, and *F* for fifty-one days, they closely resemble the early part of *B*.

Thus, the ripening of the first picking, day by day, can be foretold two and a half months in advance.

Later pickings can also be forecasted by similar means; they follow the flowering curve, with deviations due to shedding of the flowers.

growers will concede willingly, the number of bolls ripened depends on the number of flowers which open (Fig. 10). A certain proportion of these flowers are shed by the plant, the average amount under Egyptian conditions being constantly about 40 per cent. year in and year out. This proportion varies daily (Fig. 15) with the weather, and especially with the water-supply, any severe water shortage increasing the amount of shedding; while after each watering—if not excessive—the proportion of shed flowers diminishes, to show up again some fifty days later in the form of an increased rate of bolls opening.

Shedding.

It seems to have escaped notice that the shedding takes place almost *entirely* in the flower stage. The fact that a certain proportion of the sheddings (as collected from the ground below the plants) consists of bolls and buds is rather misleading. At any given moment there are enormously more buds and bolls on the plants of a field than there are flowers; in the middle of the flowering season the ratio would be about 20 buds : 1 flower : 40 bolls, simply because of the differences in the duration of each stage of development. If, therefore, we even found equal ratios of buds, bolls, and flowers, in the sheddings, this implies that 40 per cent. of the available flowers have been shed for every 1 per cent. of available bolls, and for 2 per cent. of the buds available.

The flower stage is thus extremely liable to shedding, possibly for reasons connected with the chemical side of pollination or with the greater transpiration of water from the open flower.

Apart from shedding, the only other factor which can seriously influence the yield is the size of the individual boll, and this, again, appears to be specific for each kind of cotton under definite circumstances, so that cotton-breeding for a big boll would appear to be a profitable line of research. With the opening of the flower the purpose of the present chapter

The Boll.

PLATE VIII.—AN UNHEALTHY PLANT.

Belonging to a strain which sheds its flowers very readily on any disturbance, this plant, while rather short of water, was attacked by the leaf-eating larvæ of the cotton-worm. Most of the early flowers and bolls had been shed. Note the flowering branch which projects farthest to the left, from which the three innermost bolls have all gone.

ends, since the history of the ripening fruit requires a separate chapter to itself; but it should be remembered that fresh flowers are opening day after day, for a period which is not usually less than two months, upon the scaffolding of branches (Pl. VII.), and that all these are being acted upon successively by all the factors of the environment which we have mentioned, and by others; by internal factors such as self-poisoning, and its antidotes, the whole result being moulded upon the basis of inherited capabilities for reaction.

Constructive and Modifying Factors.

It is convenient to think of these environmental factors as acting in two ways: either as "constructive," influencing the rate of construction of the scaffolding; or as "modifying," producing deformation of the results which would have been anticipated from the constructed scaffolding. Thus, sufficient scaffolding may have been laid down to produce a certain yield in a certain way, but the action of water shortage may produce shedding, attacks of insect pests may check the development of the buds, or a frost may kill the whole plant; such factors are modificatory (Pl. VIII.).

CHAPTER III

THE DEVELOPMENT OF THE BOLL

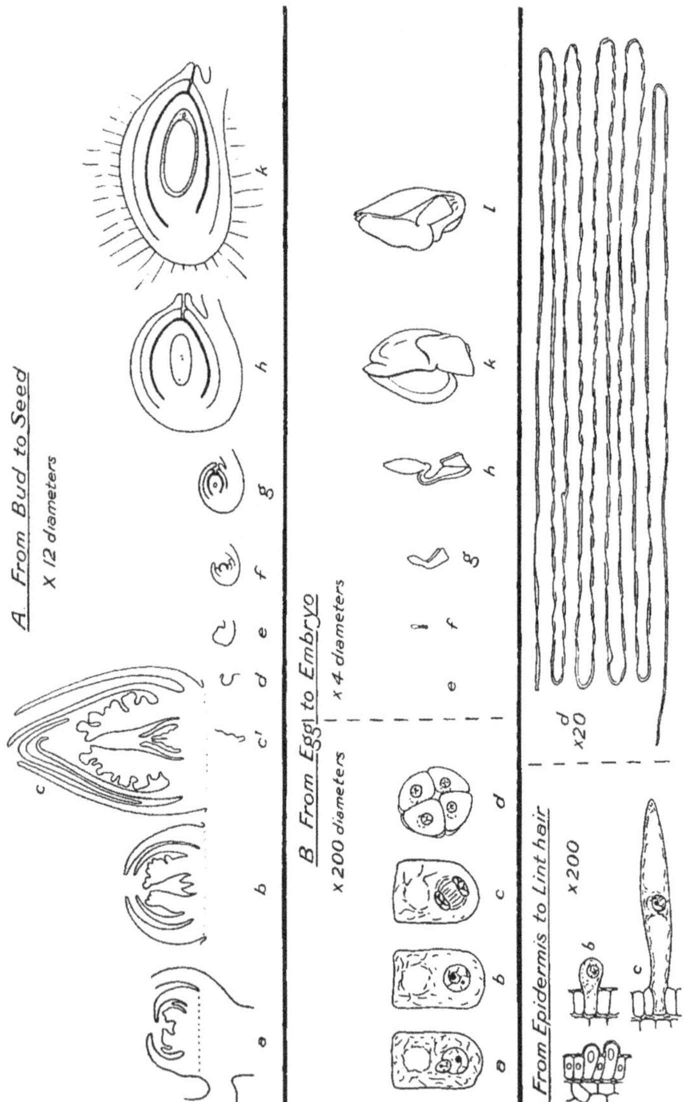
A. From Bud to Seed
X 12 diameters
a
b
c
c'
d
e
f
g
h
k
B From Egg to Embryo
x 200 diameters
x 4 diameters
a
b
c
d
e
f
g
h
k
l
C. From Epidermis to Lint hair
x 200
a
b
c
d
x20

FIG. 11.

FIG. 11.—MICROSCOPIC DETAILS.

A. FROM BUD TO SEED.

a, Flower-bud three to four weeks before flowering.
b, Flower-bud a few days later.
c, Flower-bud about two weeks before flowering.
c^1, Ovules (embryonic seeds from *c*).
d, *e*, *f*, *g*, Successive stages in the development of the ovule showing the origin of the seed-coats.
h, Ovule at time of flowering, ready for fertilization, with large central embryo sac, in which is contained the egg-cell. Same age as Day 0 in Fig. 13, p. 80.
k, Seed, corresponding to about sixth day in Fig. 13. Endosperm dotted, with embryo embedded in it near the point.

B. FROM EGG TO EMBRYO.

(*Magnified* 200 *diameters*.)

a, Fertilization of egg by male nucleus. Day 1.
b, Resting fertilized egg—*i.e.*, one-cell embryo. Day 2.
c, First division of same, to form two-cell embryo. Day 3.
d, Third division completed, forming spherical eight-celled embryo.

(*Magnified* 4 *diameters*.)

e, Embryo on fifteenth day.
f, On eighteenth day; seed-leaves visible.
g, On twenty-first day; bending.
h, On twenty-fourth day; seed-leaves folding over.
k, On twenty-seventh day; seed-leaves beginning to enwrap the rootlet.
l, Ripe embryo.

C. FROM EPIDERMIS TO LINT-HAIR.

(*Magnified* 200 *diameters*.)

a, *b*, and *c*, Epidermal cells of seed-coat budding out to form hairs.

(*Magnified* 20 *diameters*.)

d, Complete ripe lint hair, diagrammatically reproduced from actual drawing, showing relative length and diameter, and the twist.

CHAPTER III

THE DEVELOPMENT OF THE BOLL: I. STRUCTURAL

THE previous chapters have dealt with the origin of cotton, and with the development of the plants from seedlings to become the bearers of fruit. Reference can be made to the Appendix for the methods of studying the fruit itself, and the cotton formed therein.

Those readers who are not interested in the ways by which conclusions are obtained, but only in the conclusions themselves, may take it for granted in the following pages that, whenever a statement of fact is made, it results from one of two methods—either the observation on which it is based has been repeated sufficiently often to leave no reasonable doubt as to its truth, or (in the case of numerical statements) the degree of uncertainty attaching to the statement is exactly known and definable. That some statements may be qualified by reservations does not imply that they are imagined to be incorrect.

Expressions of the author's personal opinion as formed by deduction from the data are an entirely different matter. These opinions about raw cotton run counter to those generally accepted by previous writers, and, in order to afford the reader an opportunity of satisfying himself as to their validity, the rather exceptional course has been followed of presenting in the Appendix a great part of the experimental data themselves in the form of tables, instead of merely summarizing them in the diagrams.

The biologist will probably see that a number of reservations would have to be made if the arguments were carried much farther, but such reservations would be due to general causes—such as the very recent development of Limiting Factor methods in the study of Growth—and not to any fundamental uncertainty as to the particular case of the cotton fibre.

The spinner will also see that similar uncertainties are only just avoided by restriction of the scope of the work. It is the author's hope that either he or some other student may ultimately be able to write a sequel to this book, by similar examination of lint at various stages of manufacture. Even simple breaking-strain determinations before and after the processes of ginning, baling, pressing, bale-breaking, scutching, carding, drawing, combing, spinning, etc., would be well worth having, if the fate of a bale from a single field could be followed in this way, since from such a numerical basis established for an obvious characteristic it would be easier to proceed step by step to elucidation of the more subtle features, with which this book can scarcely profess to deal.

In the year 1905 a small book dealing with the microscopic structure of the cotton-plant was published by Mr. Flatters, containing some very pertinent remarks on the world's ignorance of the subject. Owing to lack of material, the subject of fertilization was not dealt with, the work having been done in England, presumably on material sent from abroad. At the same time, but independently, the present author published a paper dealing with the development of the flower from the primordial bud until a few days after its opening, and incidentally dealing with the earliest stages of the development of the lint. These two accounts both showed that the accepted story, as given by Dr. Bowman, was not reliable.

Previous Authors.

The errors of Bowman's description, which has been quoted as late as 1904, were mainly due to natural causes, this description having been published in 1881, when botany was considered to be an "inaccurate science," and our knowledge of plant and cell structure was much smaller. Unfortunately, a new edition of Bowman's work was published in 1908, which repeats almost all the original mistakes in greater detail. Monie's work, published in 1904, need not be mentioned in this connection, since it is largely based on Bowman's book of 1881, with some philosophical additions. The only critical statements on the subject, other than Flatters', are in Mr. Scott Taggart's book, which does not profess to deal with the development of the fibre. The justifiable comments made in this work provoked a reply from Monie in the 1904 edition of his book, but the effect of this reply was nullified by the fact that in the preceding lines mention had been made of "vital fluids, which are composed of a creamy-coloured oleaginous matter."

The amount of copying which has taken place may be realized by the fate of a rough sketch of the ovary which first appeared in the 1881 edition of Bowman, and has returned to England again in Matthews' work dated 1904, where it is acknowledged as being derived from a German book by Witt

To deal in detail with the various misstatements which have thus been passed from hand to hand, would take too much time, and would, moreover, be scarcely courteous to the original work of Dr. Bowman. One point alone requires notice—namely, the cause of the origination of many of these mistakes.

Cause of Former Errors.

The developing boll grows to its full external dimensions in the first half of its maturation period. Thenceforward its external appearance remains unchanged, so that unless flowers are labelled with their dates in field-crop, the microscopist will be confused by finding different stages inside bolls which appear externally to be fully developed and ready to open. The confusion thus introduced has been accentuated by the use of greenhouse plants grown in England, instead of normal plants grown in cotton-fields. The study of the abnormalities shown in the former conditions is a separate subject for research in itself. The truth of this criticism is demonstrated very well by Fig. 34 in the 1908 edition of Bowman, where five bolls are figured as ten, twenty, thirty, forty, and sixty days old, respectively. Of these, the twenty and forty day bolls are typical of what the Egyptian calls "nabroon," or a stunted condition due to premature partial obstruction of the stalk by a cork-layer, a process which carried further leads to shedding. If the very marked indication in this figure of the lines along which the boll opens may be trusted, all the bolls figured were extremely unhealthy.

Trustworthy conclusions as to details cannot be drawn in this way, and to force these abnormal plants into publicity with one hand, while writing on the extreme sensitivity of cotton to environmental influences with the other, can lead to no great progress in the mutual confidence between grower, spinner, and scientist, which is needed for the benefit of the trade.

Material.

The following account of the development of the fruit, seed, and fibre, is based upon material taken from a pure strain of Egyptian cotton, known as No. 77, producing lint which is akin to the Nubari variety of Egyptian, but rather longer and stronger. The account has been checked by examination of many other kinds of cotton, both Egyptian and American.

THE FLOWER-BUD (Fig. 11).—The first appearance of this characteristic bud, with its three-cornered cover of leafy bracts, takes place between three and four weeks before it opens into a flower. It is formed at the end of the flowering branch, and any further extension of the latter is indirect. In the case of vegetative branches the terminal bud persists, and continues to develop fresh leaves and stem-joints. In the case of flowering branches the terminal bud is differentiated into a flower-bud, and the further extension of the branch results from a new bud arising in the axil or junction of the leaf next below the flower-bud (Pl. VIII.). This new bud grows out in almost a straight continuation of the old stem, forcing the flower-bud and its stalk sideways, and itself repeats the same performance. Each new flower is therefore in reality a lateral branch from its predecessor, though the general appearance is as if each flower was an offshoot from a main flowering branch.

The flower proper begins to develop inside the three enclosing bracts at a date which can be fixed with certainty as being twenty-three days before the flower opens, in the case of No. 77 strain grown at Giza.

During these twenty-three days the various portions of the flower are developed, and their internal structure is differentiated.

The chief of these latter processes is the formation of the male and female sex-cells; but as the author has dealt with them fully elsewhere, and as the details are not immediately relevant to our present topic, a few general remarks will suffice.

The various parts of the open flower are the green ring of the calyx (which must not be confused with the three external leafy bracts), the coloured petals, the hollow brush of stamens containing the pollen, and the ovary containing the ovules. The latter develops after flowering and fertilization into the fruit, or boll, and into the seeds, respectively.

The pollen-grains each contain two male nuclei and one other nucleus. The nucleus is a definite structure found in the protoplasm of every living cell (*e.g.*, Fig. 11, *B*, *b*, and *C*, *c*), there being usually one nucleus to each of the cells which by their growth and multiplication build up the bodies of all living organisms. The function of this nucleus appears to be that of directing and controlling the behaviour of itself and of the rest of the living substance, or protoplasm, of the cell.

The transmission of inherited characters also appears to be centred mainly, if not entirely, in the nucleus of the sex-cells.

The ovules contain each a large cell in the centre (Fig. 11, *A*, *g*, *h*), in which is a single female or egg-cell, with its nucleus, accompanied by a bevy of seven other nuclei, each with its definite function. Fertilization of the egg-cell by fusion with one of the male nuclei of the pollen-grain (Fig. 11, *B*, *a*) produces a fusion nucleus, or fertilized nucleus (Fig. 11, *B*, *b*). The fertilized egg-cell thus formed is the first origination of a new plant, and on looking down the microscope tube at two fusing nuclei, whose aggregate diameter is only 0·015 millimetre, or one-fifteen-hundredth of an inch, it must be realized that the nature and origin of these two specks of protoplasm is determining irrevocably the nature of the plant which will grow from them (Fig. 11, *B*).

Of these various portions of the flower, the first to show itself in the developing bud is the calyx. The rounded end of the branch within the three tiny bracts begins to make a ring-shaped protuberance. Within this ring two other rings are formed, the outer one having an undulating edge, and its five undulations growing into the five petals. The inner ring grows up into a hollow cylinder, and the stamens arise on the outside of this cylinder as isolated protuberances, which develop into the stalked yellow or buff-coloured sacs containing the pollen-grains.

When these latter protuberances are so far advanced that the processes which will result in the formation of the pollen-grains themselves have already begun, a fourth and last set of rings arises on the end of the branch, within the staminal cylinder. These rings develop into the ovary, and thence into the boll. They are variable in number according to the kind of cotton, and are situate around the centre of the end of the branch. Each one as it grows becomes larger at the base, forming a flask-shaped body (Fig. 11, *A*, *c*); the necks of these flasks adhere to one another, and ultimately form the "style" of the flower, which protrudes through the hole in the centre of the cylindrical brush of stamens, and is destined to receive the pollen-grains which effect fertilization. The bodies of these flasks similarly adhere, and form—according to the number of flasks—an ovary, consisting of two, three, four, or five loculi, and ultimately a boll, which will open into two, three, four, or five divisions, or "locks."

On the inner wall of the body of each flask (Fig. 11, *A*, *c'*) there next appears a double row of small protuberances, from which are formed the ovules, and consequently—after fertilization—the cotton-seeds. Each of these protuberances as it enlarges forms round its base two annular swellings, first a lower one, and then another above it; these are the beginnings of the seed-coats. The embryonic seed-coats grow up over the protuberance on which they are formed, and completely enclose it in a double jacket (Fig. 11, *A*, *d*, *e*, *f*, *g*), excepting for a minute hole at the top, through which the fertilizing pollen-tube will ultimately enter. Until fertilization has been accomplished, these seed-coats are of uniform structure internally, excepting that the outermost and innermost cell layer of each can be distinguished as an epidermis. The outer epidermis of the outer coat gives rise to the cotton-fibre itself (Fig. 11, *A*, *k*; *C*, *a*, *b*, *c*; Fig. 13).

THE FLOWER (Pl. III.).—The processes so far described have taken place during twenty-three days, and when the flower opens on the morning of the twenty-fourth day it consists of three leafy bracts, about three inches long, cut into long narrow teeth (in many cottons), enclosing an inconspicuous green calyx-ring, from within which rises the bell or trumpet shaped corolla, with its brush of stamens, and the more or less protruding style. By about 9 a.m. on the day of flowering the corolla is fully expanded, and the sacs of pollen are gaping open, setting free the pollen-grains. By the evening the corolla has faded, turning in the process to pink and red colours if the fading is taking place in damp air, or simply drying up if the air is dry, with the colours almost unchanged.

Meanwhile pollen has reached the style, either by accident or brought by bees, which have visited the flower for the honey which it secretes from the nectaries between the bases of its petals. In either case the pollen may be derived from the same flower or from a foreign flower. In the former case the flower is self-fertilized; in the latter it is "crossed," and this latter event happens in 5 to 10 per cent. of the seeds ripened in an Egyptian field. Lastly, if in the latter case the pollen is derived from an identical brother-plant, such as another member of the same pure strain, the effect is the same as self-fertilization; but if—as is likely to be the case in commercial varieties—the pollen-parent is not exactly identical, the effect is to produce a hybrid embryo, and consequently a hybrid plant in the following season. The exclusion of such foreign pollen (Pl. XIII.) is the great difficulty confronting those who attempt to introduce new varieties of cotton or to improve old ones.

The action of the pollen is very like that of a parasitic fungus. The pollen-grain germinates when placed on the style or in a 2 per cent. solution of sugar, and sends out a tube, which grows between the cells of the style tissue, through the inner walls of the ovary, and enters the cavity of the ovary. There its end creeps over the surface of the ovules until it enters the small hole left by the seed-coats. Passing down this canal through the two coats, it reaches the tissues inside, and burrows through them till the end reaches into the large central cell already mentioned (Fig. 11, *A*, *h*) as containing the egg-cell and its seven attendants. There the end of the tube literally swells up and bursts, setting free the male nuclei inside the large central cell, or, as it is commonly called, the embryo-sac. One of the male nuclei fuses with the egg-cell, as already mentioned, forming the unicellular Embryo. The other fuses with two of the attendant nuclei, forming a "triple-fusion nucleus," the fate of which is to provide by its division some tissue called the "endosperm" (Fig. 11, *A*, *k*), which at first surrounds the young embryo, and is subsequently destroyed and digested by it during its growth, until only a papery layer is left in the ripe seed between the embryo and the seed-coat, along with an outer papery layer which represents the remains of the tissue in which the embryo-sac was situated. Thus the embryo-sac filled with endosperm absorbs and digests the surrounding tissue up to the seed-coats, and is itself absorbed and digested later by the embryo, till only the embryo and certain layers of the seed-coat are left alive, one complicated structure after another having been in its turn developed and sacrificed to its purpose. Lastly the seed-coat dies also.

Of the many pollen-tubes which germinate on the style, only some twenty or so can find an ovule. Those which were too late in starting, or too slow in growing down the style, also perish, and their remains may partly be traced in the walls of the young fruits; while the rest are thrown off when the style breaks away from the point of the young boll.

Some remarkable features of this race between the pollen-tubes require further study. The style of some kinds of cotton is either non-nutritious, or more probably poisonous, to the pollen of other kinds. Thus, crosses between the Indian group of cottons and the Upland or Peruvian groups do not appear to be possible. Uplands and Peruvians can, on the other hand, be artificially intercrossed with ease. Even in this case, however, if equal amounts of self and foreign pollen are placed on the style simultaneously, so that both have an equal chance, 97 per cent. of the victors will be self-tubes; so that, although Egyptian pollen can grow down an Upland style quite satisfactorily, it cannot grow so fast as the Upland's own pollen can do, and *vice versa*. If, lastly, the pollen mixture is made with pollen from the first cross between Upland and Egyptian, the percentage of wins credited to the home team falls to about 60 per cent.; the hybrid pollen is said to be more "prepotent." These facts have considerable economic bearing on the possibility of keeping cotton strains pure with fewer precautions, but it will take a great deal of tedious research to find whether they have any utility.

THE FRUIT: GENERAL (Fig. 13).—The few days immediately preceding flowering, and the day of flowering itself in particular, are extremely critical ones in the history of the fruit. For some reasons which are not yet understood, the open flower is extremely liable to "shedding" (see p. 48). The cause which provokes such shedding, with consequent complete loss of the fruit from its corner of the plant scaffolding, is usually shortage of water, either for the whole plant or for the particular fruit concerned. The shedding of older fruits is much rarer, and in all probability the disposition to shedding is connected with the presence of some chemical substance or substances formed just at the flowering stage.

The history of the fruit proper may be taken as beginning with pollination, or perhaps with fertilization of the egg-cell.

The flower having opened completely by nine or ten o'clock, is pollinated shortly afterwards. By the afternoon of the following day the egg-cell has been fertilized, and is in a resting state. The first day is thus marked by fertilization (Fig. 11, *B*, *b*).

On the third day this resting egg begins to divide into two cells (Fig. 11, *B*, *c*), which again divide, and continue to do so, till at the end of a week the embryo is visible under the microscope as a body beginning to show a heart-shaped outline, about 0·01 millimetre long, and therefore scarcely visible to the naked eye. From this its increase in size continues rapidly, till at the end of the fourth week it has completely filled the seed (Fig. 11, *B*, *l*).

Meanwhile the seed has also been enlarging, and in rather less time has attained its full and final dimensions.

The boll also reaches its full length and diameter in this period, and the lint, which began to sprout from the seed-coat on the day when the flower opened, has attained its full length.

The period of maturation, from the opening of the flower to the opening of the boll, being forty-eight days in the case of strain No. 77 at Giza, the boll thus appears to be fully grown at the end of twenty-four days. In the remaining twenty days, however, much remains to be done in the way of structural differentiation. The seed-coat hardens to protect the embryo, the embryo differentiates its internal structures ready to begin an independent existence when the seed is sown, and the lint thickens its walls for reasons which, as Mr. Flatters pointed out, have usually been ascribed to the direct benefit of Lancashire!

We have already mentioned the confusing effects which this stoppage of external growth has had upon the previous accounts of boll development, and it remains to point out that the phenomenon is not new or strange. It is, on the contrary, the usual procedure of plants first to enlarge the cell or organ, and then to differentiate inside the skeleton of delicate cell wall or tissues thus formed. That it should have taken so many years to be noticed in the case of the cotton-boll is almost incomprehensible.

We will now examine some of the details of this development, taking the boll itself first, then the seed with its components of seed-coat, endosperm, and embryo, and lastly the lint.

It should be remembered throughout that the figures and dates assigned are *averages*. In the case of any given boll they may be distorted to a definite amount, but under the conditions attained in ordinary good irrigated cultivation this distortion will not be more than about 4 per cent. either way in about half the number of bolls observed, while a distortion of 15 per cent. will be practically impossible.

Development of the Boll (Fig. 13).—The full-grown boll of No. 77 strain is 26 millimetres in diameter and 40 millimetres long, with its maximum diameter at two-thirds of the distance from the tip. About one boll in ten has only two divisions instead of three, this peculiarity of the strain being a useful mark. The ovary from which it develops is about 5 millimetres in diameter

The yellow petals having faded during the afternoon of the day of flowering, they wither and dry up during the first day of the boll history, and on the second or third day have usually fallen off, carrying with them the faded brush of stamens, and breaking away the remains of the style. This cutting off of the petals is effected by the differentiation of a layer round their base, the cells of which separate from one another instead of remaining united after division. The process of shedding of the bud, flower, or boll, is effected by the plant in the same way, the layer of special cells in this case being formed across the stalk. In both cases there is probably a reaction to some chemical stimulus, normal in the one case, abnormal in the other. Some accepted accounts of cotton inform us that the petals undergo a "peculiar rotation" on their own axis, whereby they "twist themselves completely off." This is scarcely correct.

The diameter of the ripening ovary, or boll, has meanwhile been increasing rapidly. From some 4 to 5 millimetres at flowering, it enlarges by about a millimetre of diameter each day. On the sixth day it has a diameter of 12 millimetres, or half-size; on the twelfth day 18 millimetres; and on the eighteenth day 24 millimetres. The figures with this strain are easily remembered, being six more than the number of days. After the eighteenth day the rate is rapidly decreased, and the full diameter of

26 millimetres—*i.e.*, 24 to 28—is attained by the twenty-fifth day. Henceforward there is no visible change in the external appearance of the boll until it begins to show signs of cracking along the two or three lines of dehiscence on the forty-fifth day. On the forty-eighth day it is opening and hardening, and on the fiftieth day is ready to pick.

DEVELOPMENT OF THE SEED (Fig. 13).—The ovule, when fully matured and ready for fertilization, is about 1 millimetre long. After fertilization, when it receives the designation of a seed, it increases at the following rates:

	Length in Millimetres.	Width in Millimetres.
Third day	1½	1
Sixth day	3	1½
Ninth day.. ..	4½	2
Twelfth day ..	6½	3½
Fifteenth day ..	8	5
Eighteenth day ..	9½	5½

Thenceforward the growth ceases rapidly, and 10 by 6 millimetres is the average size of the full-grown ripe seed.

THE ENDOSPERM (Fig. 11, *A*, *k*).—The embryo-sac of the ripe ovule extends over about half its length. It maintains this proportionate size as the seed begins to enlarge, and then encroaches on the neighbouring tissue. The triple fusion nucleus already mentioned does not delay its division like the egg, but by the evening of the day on which fertilization took place (the first day of development) has divided into two separate nuclei. These repeatedly divide, until by the third day there are some hundreds of nuclei arranged in a layer of protoplasm lining the wall of the embryo-sac. The large central cavity not occupied by protoplasm is filled with cell sap, consisting of water with various salts and food substances dissolved in it. At the end of the seed, where the embryo is situated (corresponding to the tip or stalk end when viewed from the outside), the nuclei form walls between themselves, cutting the protoplasm into definite cells, in the midst of which is the embryo embedded. Walling off does not take place until rather later in the other parts of the embryo-sac. By the eighteenth day the embryo-sac is about one-quarter filled with endosperm tissue, and three-quarters with cell sap, and in three or four days more the endosperm has filled the whole sac, only to be disorganized and obliterated by the embryo within a week.

THE EMBRYO (Fig. 11, *B*, and Fig. 13).—We have already mentioned the earliest origin and appearance of the embryo. To the naked eye nothing of the nature of an embryo is visible on cutting open the seed until about the fifteenth day, for, although the embryo has been growing steadily all the time, its initial dimensions are so small that even rapid cell division does not give it any great size until several days have elapsed. This makes its ultimate behaviour all the more striking. On the eighteenth day it can be seen clearly by the naked eye, in three days more it is 2½ millimetres long, and in a week more it has practically filled the seed. The second half of boll maturation is occupied, so far as the embryo is concerned, by differentiating the internal structures, with which we are not concerned.

THE SEED-COATS (Fig. 13).—The two coats of the ovule, whose origin we have described, consist of almost undifferentiated cells, all resembling one another, excepting that inner and outer epidermis can be recognized on each coat, and that vascular tissue, or veins, traverses them to provide food and water. The vascular tissue enters the seed by its stalk, runs along the side, and then breaks up at the wide butt of the seed into short distributing branches.

The two coats, though originating separately, are closely appressed to one another, and for all practical purposes form a single layer. In the ripe seed this double jacket of delicate thin-walled cells has been converted into a horny protective envelope, consisting of the following structures from the outside inwards: There is an epidermis (between the cells of which the lint arises), an outer pigment layer, a hard "crystal layer" (*Krystallschicht* of Weisner), then a very thick horny layer of palisade-like cells, followed internally by the inner pigment layer, to which succeed the two papery remnants already mentioned as being derived from the endosperm, etc.

Though their thickness becomes greater through increased cell size and division, the coats show no marked change until the twelfth day of boll development, excepting for the growth of the lint and fuzz hairs.

On the twelfth day the outer epidermis of the *inner* seed-coat (corresponding to the layer from which the lint arises in the outer seed-coat) begins to increase the size of its constituent cells.

By the fifteenth day these cells have undergone marked

elongation in a radial direction, and their nuclei show up very distinctly, scattered about all parts of the "palisade" layer which is formed by this elongation.

By the eighteenth day the nuclei of the "palisade" layer have taken up their position at the outward end of each palisade cell.

By the twenty-first day the inner end of the palisade cells has begun to thicken its walls, so that this layer, when viewed in section, consists of one-half thick-walled, clear, translucent, stony tissue, streaked by the original cell walls, and the outer half of the zone alone contains the nuclei and granular protoplasm, with the same walls still relatively thin.

The epidermis continues to increase the depth of its constituent cells for another week, and the palisade layer builds more and more thick cell wall on the inner end of each of its own cells, thereby forcing its nuclei farther and farther from the centre of the seed, so that some three-quarters of the palisade zone consists of translucent stony cell walls.

About the twenty-seventh day, when the seed-coats have reached the state of development described, a marked chemical change takes place in the pigment layers, which has not yet been investigated, but which is of considerable interest because it denotes the beginning of the second stage of boll maturation, during which the lint is given its strength. The change shows up in material which has been preserved in a mixture of acetic acid and alcohol. All the bolls from the first stage "pickle" to a green colour, which, of course, fades to brown on exposure to light. From the twenty-seventh to forty-fifth day, however, the pickle thus made is at first pink, and then bright red. The colour is probably connected with the development of pigment in the seed-coat. When the boll is beginning to crack, the pickle is brown.

By the thirty-third day the palisade layer would appear to have reached its limit of possible extension by the method of construction practised thus far, and it finishes by putting down a certain amount of thick cell wall externally to its nuclei, so that in the ripe seed the palisade layer, which comprises half the total thickness of the seed-coat, shows a granular zone running through it at about one-third of the way inwards. This zone is the dead remains of the nuclei and protoplasm, which thus constructed and sealed their own living tomb.

At the same stage the epidermis has finished enlarging, and has begun to thicken its walls.

Subsequent changes are relatively uninteresting. The inner epidermis of the outer seed-coat thickens its walls without marked alteration otherwise, and the cells which separate it from the outer epidermis are disorganized to form the outer pigment layer of the ripe seed.

All the cells lying within the palisade layer are similarly disorganized, forming the second pigment layer, the inner epidermis of the inner coat disappearing along with them in marked contrast to the enormous palisade layer formed from the outer epidermis.

It has seemed worth while to give the details of seed-coat development at some length, because of the casual assistance which they might provide to persons wishing to ascertain the age or state of ripeness of odd samples of cotton-seed or of seed cotton which might have to be examined. An entomologist, for example, could easily use them for dating the attack of a boll-worm. If used for such purposes, however, it would be advisable to take a check series when possible, as the times given would necessarily vary to some extent from country to country, and even within the same variety. Thus, at Giza, in 1913, while Strain No. 77 had a maturation period of forty-eight days, the same period on the same land for "Domains Afifi" was fifty-one days.

Development of the Lint (Fig. 11, *C*, and Fig. 13).—Whatever refinements we may ultimately be able to introduce into our knowledge of lint development, so as to explain the origin of subtle differences which only the expert can detect, between samples of the same uniformity, length, breaking strain, diameter, and colour, it is now clear that the progress of inquiry has been hampered in the past by unnecessary and mystical elaborations of what is actually a most simple story.

The origination of the fibre is quite independent of fertilization, and also of pollination. Flowers picked at noon on the day of flowering show well-defined lint hairs, about as long as they are broad (Fig. 11, *C*, *b*). The lint hair itself is a simple unicellular hair, formed by the outward extension of the external wall of a single outer epidermal cell of the outer seed-coat. The full diameter of the hair is attained almost at once, when it is only $\frac{1}{10}$ millimetre in length (Fig. 11, *C*, *c*), while its length continues to increase until the twenty-fifth day of development, after which (Fig. 13) its wall begins to thicken, giving strength to the lint. This thickening is not uniform, but leaves simple pits in the wall (Fig. 12), set obliquely, and the closure of these pits when the wall dries after the boll opens gives twist to the fibre. The uninucleate cell-contents remain alive until the boll begins to open, when they die through desiccation.

The ordinary epidermal cell of the outer ovule coat has a thick basal wall, separating it from the subepidermal layer, thinner side-walls, a thin cuticle covering the outer wall and dipping slightly between the side-walls, with a nucleus which is about one-fifth of the length of the cell, and small sap cavities in the protoplasm.

The outer wall of some of these cells bulges, their protoplasm becoming densely granular in the protuberance, while the nucleus moves up closely behind these granules. The swelling enlarges to about twice the diameter of the original cell, and the nucleus passes out into it with all the chromatin retracted into a large, deeply-staining, central nucleolus. The nucleus keeps at a short distance behind the tip of the swelling, or hair, which continues to elongate at the rate of about 1 millimetre a day, which may be expressed in terms of diameter as about twice the diameter added to the length on the average every hour. It is almost certain that in the case of Egyptian cotton this elongation is not continuous, but is intermitted during sunshine in the same way as the growth of the branches or stem.

The nucleus at a later stage appears to settle near the centre of the fibre, about one-third of the way from the tip.

The cell wall remains extremely thin for the first three weeks, and the cuticle which still covers it can scarcely be differentiated, unless the wall has been swollen by Schweitzer's reagent, when (being unaffected by the ammoniacal copper

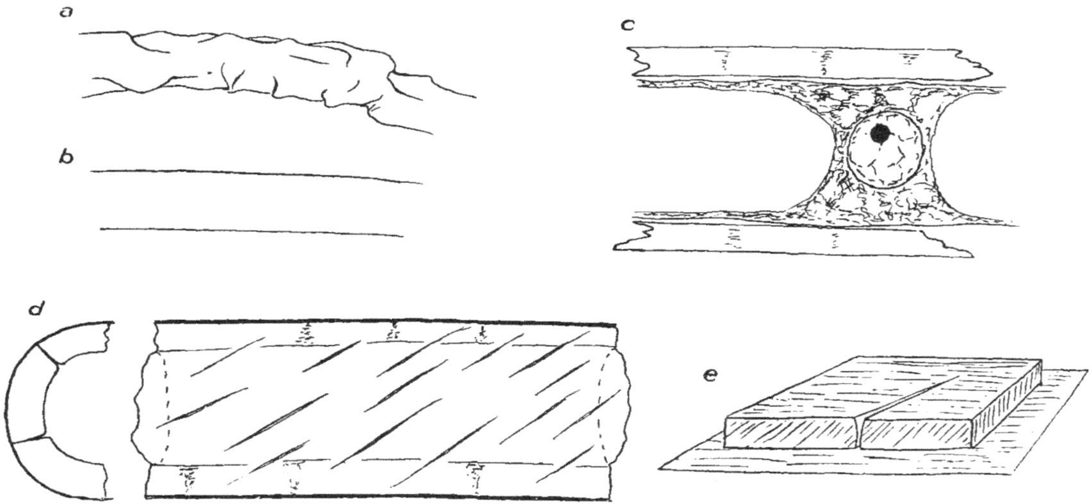

FIG. 12.—PITS.

a, Lint hair maltreated on eighteenth day, showing collapse of primary, unthickened wall.
b, Same treatment as (*a*) on twenty-fourth day, showing increased strength when even slightly thickened.
c, Longitudinal section of lint hair on thirty-sixth day, showing nucleus (one to each hair), protoplasm, thickened wall, and sap cavities. The protoplasm is slightly withdrawn from the wall; when alive, it is pressed against the wall by osmotic pressure of the cell sap.
d, Wall only, upper half (and diagrammatic end view), showing the pits in plan. Indications of them are also seen in the thickness of the wall as vague markings.
e, Diagram of half a pit. The secondary thickening is resting on a piece of primary wall, which the pits do *not* pierce.

oxide) it causes the familiar beaded appearance, the cellulose of the wall swelling through torn places in the cuticle.

The actual length to which the hair attains depends in the first place on the inherited nature of the plant which bears it, as discussed in a former chapter, and in the second place on the environmental circumstances during some ten days only, as discussed in Chapter IV. Whatever may have happened to the plant before or after these ten days cannot affect the length, exclusive only of the self-poisoning phenomena of Senescence. These latter probably enter to some extent into the life-history of every hair, but more research is needed upon them. One obvious utility of such research should be apparent, in that if senescence is the cause which checks the growth of the fibre at a certain length, it might be possible to obviate it, and produce lint of any length desired !

In the case of a batch of flowers marked on July 10, 1913, and picked at regular intervals till maturity, on No. 77 strain, the final mean length attained by all the fruits left to ripen was 30 millimetres (determined by seed-combing). The flowers from a few days before and after also gave the same length, so that the environmental circumstances were obviously acting fairly uniformly at the time. The mean length per day in the developing bolls was as follows:

Day :	3rd.	6th.	9th.	12th.	15th.	18th.	21st.	24th.	Ripe.
Length in millimetres ..	$\frac{1}{4}$	2	6	11	16	23	29	30	30

Some bolls and seeds stopped before reaching 30 millimetres, and some extended beyond it, the extreme lengths ultimately attained by sixty seeds which were measured when ripe being 24 and 35 millimetres, twenty of these being 29, 30, or 31 millimetres in length. These deviations from the mean are, of course, due to accidental circumstances, such as a check in growth from any cause, affecting the boll locally or the plant generally.

It should be noticed that the rate of growth is at its maximum somewhere near the *fifteenth day*, and is slow at the beginning. This should be borne in mind, as it will recur in the next chapter. Obviously, any cause affecting lint length will have most influence if it is acting on a boll which is fifteen days old.

Up to the twelfth day the delicate lint is firmly attached to the seed, and can only be torn away with some difficulty. By the fifteenth day, however, a noticeable change has taken place, and the lint can be stripped off with great ease during the next fortnight, leaving a smooth, shiny seed-coat behind. The subsequent firmer adhesion of the lint is due to the thickening of the epidermal cell walls, but it never again is so firmly held as during these first two weeks.

The first signs of thickening of the lint are apparent on the twenty-first day, though no visible increase in wall thickness can be seen. This first sign is a very simple one. Material which has been pickled in alcohol dries very quickly, especially when single hairs are removed and held in the air. Hairs thus treated from material earlier than the twenty-first day are contorted in all directions as they dry up, but hairs of the twenty-four day twist up as they dry. By the twenty-seventh day the wall is visibly thickened with secondary deposits of cellulose on the interior of the primary cellulose wall and its cuticle. This accession of material continues until the boll is about to crack, but the most rapid increase is noted about the *thirty-sixth* to *thirty-ninth days*.

Meanwhile, by the thirty-third day it is easy to distinguish the simple pits in the wall (Fig. 12). These pits are common in many kinds of vegetable cell-wall, and are not in any way peculiar to cotton. They are of about the same length as the diameter of the cell, and are set at an angle of about 30 degrees to the axis of the hair.

They are not at all easy to recognize; ordinary illumination with a good microscope does not display them, owing to the translucency of the cell wall and the background of granular protoplasm. A much higher and better magnification (*e.g.*, Zeiss compensating ocular 6, and 3 millimetre apochromatic objective) is needed than their size would appear to warrant, together with "critical" illumination. It should also be noticed that they cannot be seen in ripe hairs, having been obliterated by the twist, nor are they distinct in old pickled material; but they cannot be missed in the unripe, untwisted hairs, given a good microscope, new material, and correct illumination.

That they have thus far escaped notice is probably due to the fact that the observers who had the microscopes had not the material, and *vice versa*. The discovery of them renders unnecessary a great deal of speculative philosophy which has been accumulated round the subject of twist. Given such pits, the fibre must twist when it dries, unless the wall has been thickened so much as to obliterate the central cavity almost entirely. Whether the twist is right or left handed in any part of the fibre is determined by whether the pits are laid down with an inclination in one direction or the other. This direction appears to be accidental. It might be well if someone would re-examine the useless wild-cottons, which have weak twist or none, on these lines, and see if, as is probable, the pits are set too obliquely or too transversely to effect a good convolution of the hairs. It should be noticed that the twisting of the cell-wall may take place completely before collapse begins. Unripe lint kept in pickle behaves thus. The subsequent collapse which renders the convolutions visible when the boll opens can thus take place without further twisting.

The actual mode of thickening is open to further investigation; a well-thickened wall is about 0·004 millimetre thick, or about one-six-thousandth of an inch. This wall is probably composed of concentric layers, laid down during the active growth of each successive night, and numbering about twenty-five in all, of which a dozen should be of appreciable thickness. The daily or nightly layers would thus at the most be about 0·0004 millimetre in depth, so that their resolution by the microscope is highly improbable, without some previous treatment Having been led to consider the existence of such concentric layers as probable, the author made several attempts to see if they could be recognized, but without success. Indications are sometimes seen when a section across a fibre has been torn slightly by the razor in cutting. If they exist, it is possible that some of the finer details of grading may be concerned with their arrangement, and the

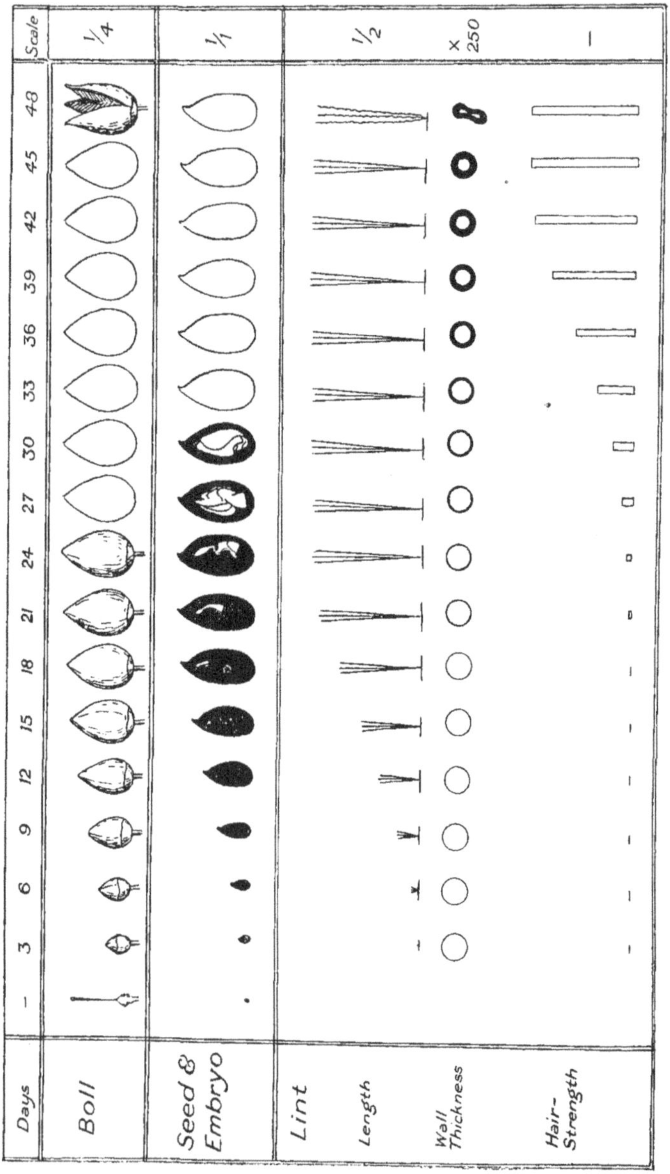
Days
Boll
Seed & Embryo
Lint
Length
Wall Thickness
Hair-Strength
Scale
1/4
1/1
1/2
× 250
1
1 3 6 9 12 15 18 21 24 27 30 33 36 39 42 45 48

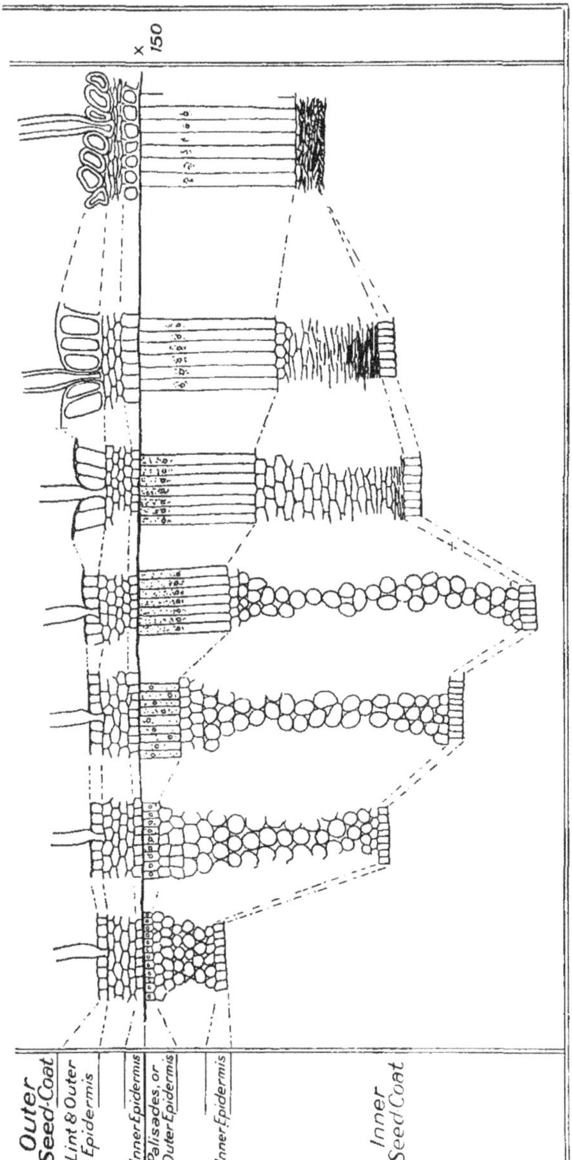

FIG. 13.—STRUCTURAL DEVELOPMENT OF THE BOLL.

Diagram drawn strictly to scale, showing the developing boll every three days from flowering to ripeness, with microscopic drawings of the development of the seed coat.

finer properties of cotton might be partly due to regularity of concentric structure, and to alternation of denser and looser layers of cellulose, analogous with the grain of timber.

Mr. Scott Taggart has raised a reasonable objection to the view that the fibre twists upon itself, but it is doubtful whether the objection is valid. He assumes that the hairs would knot themselves together inextricably if this were the case. As a matter of fact, they do knot themselves up a great deal, and now we recognize that the cell does not collapse until the boll is opening, we can allow them much more space to avoid one another than when it was believed that the twist was put in while the boll was still closed. A simple feature which seems to have escaped notice in this connection is the expansion of the seed-cotton when the boll opens. This is simply due to the twisting of the individual fibres, and when any fibre finds its freedom to twist opposed it necessarily pushes away its outermost neighbour, who is on the line of least resistance. The summation of all these little efforts expands the seed-cotton into a fluffy "lock." Very thin-walled cotton, which cannot exert a very powerful twist on drying, ripens silky locks of seed-cotton, in which the fibres run more parallel than in a more robust kind. Taking into consideration the fact that a fibre removed from the boll during the latter half of maturation does twist when dried, the evidence is probably against Mr Scott Taggart's hypothesis of collapse without twisting, as a whole; but it is not improbable that the twist takes place, as in pickled material, before the collapse which reveals the convolutions (p. 79).

The density of lint on the boll is determined when the

lint first originates by the protrusion of individual epidermal cells. There does not appear to be any further growth of epidermal cells into lint hairs after this first day, in spite of accepted statements to the contrary. The density of the lint on a given area of seed-coat should thus, other things being equal, depend on the circumstances of the environment on the day when the flower opens. We shall see in the next chapter that there is some indication of this being the case.

FUZZ.—The hairs of the fuzz are distinguishable from those of the lint by their much greater diameter, even in the earliest stages of their development. They are as a rule about twice the diameter of a lint hair, or more. They arise in much the same way, at the same time, and from the same layer of cells. We discussed some of the interesting features of this velvety covering in an early chapter, and, except by showing the markedly greater size of the hairs, the microscope has thrown no further light on its significance, nor on its evolutionary relations to the lint.

We have now traced the main outlines of the details involved in the development of an average cotton boll. A tabular statement of the dates and corresponding stages in its life is appended, by the help of which, and of the diagrams (Fig. 13), it should not be too troublesome for the non-botanist to recapitulate the story.

We now pass to a more living account of the development, in which the play and elasticity of the structures we have described are demonstrated under the ebb and flow of environmental conditions.

Day.	Boll. Diameter (m.m.).	Seed. Length (mm.).	Seed. Breadth (mm.).	Seed. Embryo.	Seed. Endosperm.	Seed. Outer Seed-Coat.
0	6	—	—	—	—	Lint sprouting
3	9	1½	1	1 cell	Layer lining embryo-sac	—
6	12	2½	1½	100 cells	Layer thickening	—
9	15½	4½	2	Heart-shaped	Embryo deeply embedded	—
12	18	6½	3½	Root and seed-leaves clearly defined	—	—
15	21	8	5	Visible to naked eye	Embryo-sac one quarter filled	—
18	24	10	6	1 mm. long	Embryo-sac completely filled	—
21	26	10	6	2½ mm. long	Quarter obliterated by embryo	Outer epidermis enlarging
24	26	10	6	5 mm. long	Half obliterated	—
27	26	10	6	7 mm. long	Nearly obliterated	Outer epidermis deeper
30	26	10	6	Full size	Obliterated	—
33	26	10	6	,,	—	Outer epidermal walls thickened
36	26	10	6	,,	—	Inner epidermis thickening
39	26	10	6	,,	—	—
42	26	10	6	,,	—	—
45	26	10	6	,,	—	Intermediate layer disintegrated
48	26	10	6	,,	Papery layer only	—
51	26	10	6	,,	—	—

OF BOLL DEVELOPMENT.

Seed.	Lint.		Notes.
Inner Seed-Coat.	Length (mm.).	Thickness.	
—	—	—	Flower open
—	¼	Cuticle and primary cellulose only	Lint firmly adherent; corolla shed
—	2	,,	—
—	6	,,	—
Outer epidermis altering to palisades	11	,,	Lint still firmly adherent
Palisade definite, nuclei scattered	16	,,	Lint easily stripped from seed
Palisade nuclei at outer margin	23	,,	—
Half palisades sclerotized	29	Secondary thickening begins	Lint twists on drying
—	30	Thickening scarcely visible	—
Three-quarters of palisades sclerotized	30	Thickening visible	Pink pickle with acetic alcohol
—	30	Pits visible	Red pickle
Outer margin of palisades sclerotized	30	—	Lint less easily stripped
—	30	Maximum increment of thickness	—
Inner layers disintegrating	30	—	—
—	30	—	—
—	30	Thickening practically ceased	Brown pickle; nuclei still visible in lint; boll scarcely cracking
—	30	—	Twisted fibres frequent; boll opening and hardening
—	30	—	Boll ready to pick; some dying protoplasm still visible in some lint hairs

CHAPTER IV

DEVELOPMENT OF THE BOLL: II. ENVIRONMENTAL INFLUENCES

By studying in the previous chapter the stages of the development of the boll in one pure strain, we reduced the problem to its simplest terms, taking from this strain a batch of flowers which all opened in the same plot on the same day, and therefore went through experiences as nearly uniform as was possible. The presentment thus obtained is, nevertheless, correct in its general outlines for other kinds of cotton.

Before we can proceed further to some preliminary attempts at analyzing the commercial cottons, a further examination is needed, in which the Environment is allowed to change. The accounts given in this chapter are taken from data obtained with the same strain—No. 77—so as to avoid confusion, and also because it is with this strain only that sufficiently comprehensive data have been garnered. As in the previous chapter, however, the results are generally applicable, with slight modifications, to other kinds and environments.

Then, in Chapter V. we shall consider the effects of altering another component of the problem as well—namely, the Constitution of the plants cultivated.

There are many ways in which the environment of plants can be altered, but the simplest way of all is to make no attempt to control the surroundings, but simply to record the changes which naturally take place. This method is almost ideal for the study of developing cotton bolls in an irrigated and rainless country. Each set of bolls opening on each day will have passed through a slightly different series of experiences from those opening on the day before or on the day following. By arranging to obtain continuous observations day by day throughout the season, we can watch the effect of any particular environmental effect upon bolls of every age. Since the understanding of the following results depends on the realization of this point, it may be dealt with a little more fully (see also Appendix I.).

Method of Study.

We saw in the previous chapter some strong reasons for concluding that any marked alteration of the environment on a certain day would affect developing bolls of various ages according to the particular structural developments which were progressing inside them at the time. Consequently, if we apply the most severe modification possible, by killing the plant outright, we shall not affect the length of the lint in the old bolls, but we shall prevent it from becoming any longer in the young ones. Similarly, since the thickness of the lint hair wall is not laid down until the boll is halfway through its maturation, we shall check any further increase in thickness.

Age of Boll.

Less severe modifications of the environment, such as

water shortage, would presumably act in a proportionately less severe way. Thus, bolls which were young when deprived of water would not make lint of the full length, but the lint might be subsequently thickened normally if normal water-supply were restored. Conversely, bolls which were three-quarters grown at the same time would not thicken their lint normally, but the length of the lint, having already been established under preceding normal conditions, would be normal and unaffected.

We will now proceed to see how far evidence obtained in this way at Giza will carry us.

Material Studied.

Two complete series of data have been obtained. The first covers sixty days in succession during 1912, on a plot of wide-sown No. 77, which was purposely subjected to severe water shortage; the second covers ninety successive days in 1913, on a group of plants of the same strain, growing in field crop conditions on excellent land, with cultivation as nearly perfect as it could be, and producing a crop from this particular plot of roughly 700 pounds of lint to the acre (if allowance is made for some 150 pounds of lint damaged by a very severe boll-worm attack). It will be seen later that the attack of this pest made no difference to the behaviour of the lint, so long as lint from obviously damaged locks was excluded from examination.

One difficulty presents itself—the dating of the bolls. Since there is a definite amount of fluctuation in the length of the maturation period, there must be uncertainty as to the exact stage of development of the old bolls if we group them by the date of flowering, and conversely there will be uncertainty respecting the young bolls if we group them by the date of boll-opening. For very accurate purposes the flowers should be dated, and only those bolls which had ripened from them at the average maturation interval should be picked. This would mean labelling about five times as many flowers as were actually used, apart from those lost through normal shedding.

In these two series the author employed flower-labelling for the first series, and daily picking for the second. The first series was more closely directed to the study of lint length, and the second to the study of lint strength. The second method, moreover, accords with the actual field practice.

That a real difference, though a slight one, exists as between the two methods may be seen by examining the following table, which shows the "variability" of lint length from bolls which were all of the same nominal age, as determined by the two methods, taking sixty-three seeds in each case, and measuring the length on each in millimetres.

Length in Mm.:	28.	29.	30.	31.	32.	33.	34.	35.	36
Dated by flowering	1	1	2	8	21	15	13	5	3
Dated by boll-opening ..	1	6	9	7	12	17	5	5	1

The lengths are much more irregular in the second case, simply because we have included (under the same nominal age-designation) bolls which were not of the same age when the lint length was being determined, whereas in the first case our grouping is not likely to be more than a day wrong either way in this respect.

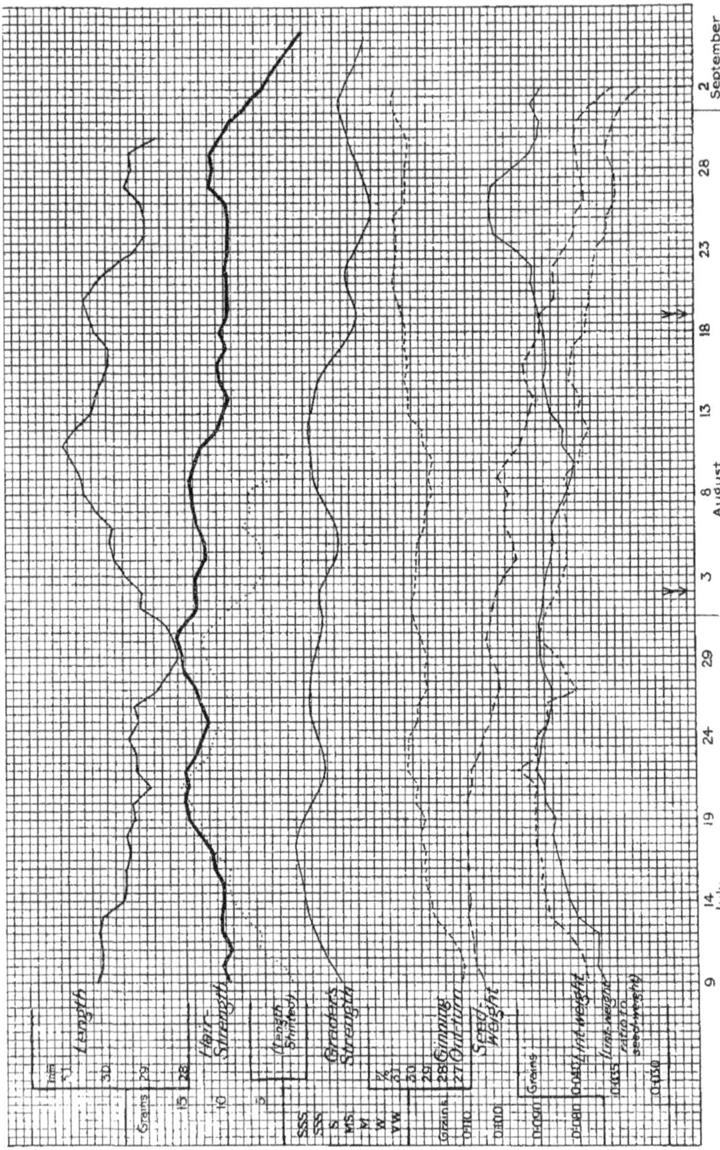

Length
Hair-Strength
Length Shifted
Grader's Strength
Ginning Out-turn
Seed-weight
Lint-weight
Lint-weight ratio to seed-weight
9
14
July
19
24
29
3
8
August
13
18
23
28
2
September

FIG. 14.

FIG. 14.—DAILY FLOWERS, 1912.

Diagram showing the changes in the principal characteristics of the seed-cotton from day to day, in samples ripened from flowers opening on sixty successive days; from left to right.

The curves from above downward (on the left side) represent:

Lint length, *thin line.*

Hair strength, *thick line.*

Lint length shifted backwards twenty-one days to synchronize the action of the environment upon length, with its action upon strength which—in any one boll—is determined twenty-one days later (see Fig. 13, pp. 80-81), *dotted line.*

"Strength," as estimated by experts' grading, *smooth curve.*

Ginning out-turn, *small dashes.*

Seed weight, *longer dashes.*

Lint weight per seed, *dots and dashes.*

Lint weight as a percentage of the seed weight, *continuous line* (crossing the two preceding curves).

The chief feature of this diagram is the resemblance of the dotted line to the thick one.

Arrows at the base denote the dates on which water was given by irrigation.

Dated Flowers Experiment of 1912.

The plot of No. 77 cotton was wide-sown with plants left standing singly on ridges 1 metre apart, the plants being separated by 1½ metres on the ridge. Cultivation was normal until early June, when irrigation was deliberately delayed, and further irrigation was then withheld from June 19 to August 2, instead of being given on July 10 as well.

Twenty flowers were labelled daily from June 7 to September 1, and were picked as they ripened. There were no insect pests, with the exception of abundant "stainer-bug" and a moderate amount of boll-worm in the last bolls.

Statistics.

The actual numerical determinations made are all given in Appendix II., Table I., which shows how the length and breaking strain of the lint, the ginning out-turn, and the weight of a single seed, were actually determined. The calculated weight of lint on a single seed, and the same figure reduced to a standard seed weight of 0·1 gramme, are also included. For the general reader, however, the main interest centres in the final figures for length and strength expressed as five-day means (Appendix II., Table II.), and in the presentment of these and other five-day means in the diagram (Fig. 14).

Soil-Water.

On this diagram are marked the dates of irrigation, and we could if necessary include all the other factors of the environment, such as temperature, wind, evaporation, etc. To do so would complicate matters unnecessarily, as it is clear that soil-water is the chief factor involved.

This influence of soil-water is characteristic of the Egyptian crop during the ripening of the bolls. In the middle of June the climatic control, which has until then been the main factor of the environment (acting chiefly through night temperature, as we have formerly mentioned), is rapidly lost, with the increasing size and evaporation of the plants, and thenceforward the chief need is the maintenance of sufficient moisture around the roots.

If we now examine the curves showing the changes in these various characteristics, such as lint length and strength, from one five-day mean to the next, the first feature which catches our attention is the suddenness with which the changes take place. The length of the lint rises steadily from 29·1 millimetres on August 1 to 30·9 millimetres on August 10, or nearly $\frac{1}{12}$ inch in ten days; and this does not give the full magnitude of the change, since the calculation of five-day means tends to smooth out these differences. The true change between July 29 and August 12 is nearly 4 millimetres, or $\frac{1}{6}$ inch, which in itself is sufficient to change the commercial classification of the lint produced.

Alterations from Day to Day

These sudden changes are shown by all the observed characters, and it should at once be obvious that, when we speak of the properties of a bale of lint ginned from the field crop of even a pure strain of cotton—much less from a commercial variety—we are speaking in averages. That the pickings from different parts of the same field may be different has long been recognized, as also the fact that

Average Properties of a Bale.

different pickings have different values; but the extreme rapidity with which the properties of the cotton may change, so that the picking of one day may be capable of differentiation from those of the preceding and following days, has not been previously demonstrated.

Since the particular series under discussion was not grown under true field crop conditions, we will postpone further comment on this point.

The next notable feature of these records is the entire lack of apparent connection between the various curves.

Independence of Properties.

Taking the most important cases from the economic view-point, Length and Breaking strain of the lint, we find the length falling steadily while the strength is rising unevenly, so that the pickings which ripen from the flowers of July 29 are both the strongest and the shortest of the whole series. Thenceforward there is a general rise in length and fall in strength, so that the flowers of August 11 ripen into lint which is the longest of the series, but is about 20 per cent. weaker than that of July 29.

A first casual inspection of the length and breaking strain diagram might thus lead us to the conclusion that

Possible Relations of Length and Strength.

a ripening boll had the choice between one of two careers, in so far as its lint was concerned: it could either become strong or it could become long, but it could not attain to both at once. A further postulate of some external change which would cause the bolls to turn their attention in one direction or the other would complete a theory of cotton development which would not be far remote from the generally accepted opinion of the present day.

Further examination would destroy this conclusion, for it would then be noticed that the two curves do not run exactly counter to one another, but that there are occasional minor rises and falls which are the same in both. These might be due to two causes: either the hypothesis of antagonistic development is wrong, and the general antagonism of the two curves is mere accident; or the methods by which these lengths and strengths are determined are not sufficiently accurate, and the minor rises and falls are of no significance.

Statistical Methods.

Here comes in the utility of modern statistical methods; by their aid we can give a numerical expression to the chances of inaccuracy for these points which compose the curves. It is not necessary to go into details of the way in which these "measures of inaccuracy" are derived, but the result in the case of these two curves is as follows: For any point in the length curve the chances are even that it is not more than 1 per cent. out of the position which it would occupy were infinite pains and repetition used in its exact determination, while it is highly improbable that it should be more than 3 per cent. out of place. For the curve of breaking strain the even chance is 1·5 per cent., and extreme improbability at 5 per cent.

We now take the rise in strength, which culminates on August 9 (August 5 to 12); we find that the rise is 10 per cent. If only August 5 and August 9 were available, this difference might just be explicable by the summation of two extremely improbable occurrences. There are, however, some eight days involved, all in regular sequence; since the addition of each extra observation decreases the probable uncertainty of the mean of all the observations according to a definite law, it follows that this rise in strength, which culminates on August 9, is not due to deficient precision in the methods used, but was a real rise which the plants actually experienced.

We can now go back to the other alternative, assume that the general antagonism between length and breaking strain in this series was mere accident, and see whether our knowledge of the structural development will help. We decided in the previous chapter that the most rapid increase in lint length took place about the fifteenth day, while in the thickness of the lint hair wall it took place at about the thirty-sixth to the thirty-ninth day

Length degraded by Water Shortage.

We will first consider the fortunes of some flowers opening after July 29, remembering that this date was near the end of a long period of water shortage, when the plants were scarcely retaining any of their flowers, and were probably more or less poisoned or senescent Flowers opening on August 29 were in their fourth day of development when the plot was irrigated, which we have seen in the last chapter would imply that their lint was about 10 millimetres long; they went on to maturity, and produced lint of 28·2 millimetres length. Flowers which opened after this day, up to those which opened on August 11, were of course younger when the water was given; those which opened on August 11 were then young buds which had only just begun to form their pollen-grains, for example. The younger these buds were when the water was given, the more time they had to recover from the poisoning effects of water-shortage in the hot, dry climate of Egypt, and the more opportunity they had to decompose the poisonous substances which are believed to be formed in the cells. As the amount of this poisonous substance decreased in successive flowers, the lint grew up more nearly to its full length, but—as we shall see in the Daily Picking Series of 1913—did not reach it before the soil began to dry up again, and from August 11 the length therefore began to fall.

The full significance of this poisoning effect has yet to be worked out, and it would seem that cotton-lint is most suitable material for the purpose. Herein consists one of the principal utilities of pure strains: If they do not produce the product which we know they are capable of giving us, we can recognize the fact at once, and can search for the cause. The whole behaviour of No. 77 in this Dated Flower Series was rather that of a good strain struggling under adversity; it was prevented from reaching its normal behaviour by the poisoning effects brought about through water-shortage.

Utility of Pure Strains.

The main feature of these curves remains to be discussed. We have seen that they represent the behaviour of a cotton-plant under severely adverse conditions, and that there is no significant connection between the properties of length and strength in the lint from any given boll. In spite of this the movements of one curve can be used to forecast the other, so that, if we know the breaking strain

Length and Strength due to the Same Causes.

of the lint from bolls opening on a certain day, we can prophesy what will be the length of the lint in later bolls.

Let us take the case of a flower opening on August 11, which we have seen already was a day giving the young fruit a chance to produce lint of good length. This would probably be due to causes—in addition to the recovery from poisoning—acting when it was about fifteen days old, as the previous chapter indicated, and if we refer to the diagram we shall find that when this boll was eight days old the land was watered (on August 19). Thus on August 19 to 25 we know that certain environmental conditions existed which were favourable for lint length development.

It remains to see what effect these same conditions produced on the thickness of the lint hair wall, affecting the breaking strain. Such bolls as would receive the most benefit from these optimal circumstances—if such were capable of acting equally on thickness as well as on length—would presumably be about thirty-eight days old at the time. Bolls which were thirty-eight days old about August 24 would have opened as dated flowers about July 20. The flowers which opened about July 20 are seen—in the diagram—to have had nearly the strongest lint hairs in the series.

We can test the matter for every date examined by shifting not merely the flowers of July 20 and August 11 into superposition, but by moving the whole curve of lint length back through an interval of about twenty-three days, as has been done in Fig. 14. *The two curves are the same,* when duly synchronized, excepting for alterations in general slope, which are due to the self-poisoning effect.

This identity of the two curves fits in perfectly with the microscopic evidence, and we shall see in the next series that it was no mere accident of the season which brought it about, nor any abnormality of the plots, but that it is even more marked and definite in good cultivation than in experimental modifications.

Before considering the behaviour of the other characters it may be worth while to comment on one detail of the preceding pages. The author has shown elsewhere that it is possible to depict "good cultivation" graphically, for under a given set of climatic circumstances a given kind of cotton should flower and fruit at certain definite rates, and if these rates are not attained cultivation has been defective. Following on from this, and from the remarks on poisoning from water-shortage just made, and supplementing them by comparison with the series of dated bolls in field cultivation, it becomes possible to define the object of good cotton cultivation as a fight against self-poisoning, or senescence, or autotoxy.

Seed-weight is a feature which we can hardly expect to dissolve into its components. To some extent it is determined by the size of the seed, which is settled at the same time as lint length, and should therefore fluctuate with it. The mere size, however, is not everything, and all the subsequent changes which the embryo and seed-coats undergo must each leave its mark upon the weight finally attained. We

The Seed.

might expect that the seed weight would roughly follow the mean between the daily changes in lint length and breaking strain, and this seems to be the case. Senescence effects are more markedly shown by the seed, with its massive cell structure, than by the more or less isolated simple cells of the lint, so that the mean seed weight in this series degrades steadily towards the autumn.

The seed itself is of comparatively little interest to those who have to deal with raw cotton, but to the grower and to the owner of the ginning factories the out-turn in ginning is a matter of considerable importance. The cause of the very definite seasonal and geographical variations which take place in this respect has never yet been explained; differences between different kinds of cotton have been partly traced to their source, but the causes of fluctuation in ginning out-turn have long been mysterious. Part of this is due to the difficult way in which the ratio is expressed as "lint obtained from lint plus seed." If we take the data for ginning out-turn and convert them into absolute measurements as "seed weight" and "lint weight per seed," we shall find that they are easier to handle.

Ginning Out-turn.

Lint/Seed.

In the first place, we note that towards the end of the season, when the lint is becoming short and weak, the out-turn at the gin rises to its maximum. This is not the common experience of field crop, but is presumably the accidental outcome of our abnormal treatment of the plot in question. It immediately causes one to suspect the

Out-turn and Quality not necessarily connected.

truth of the belief that there is a necessary connection between high out-turn and good lint; probably it is a matter of accident that the circumstances which produce high out-turn do also produce good lint under field conditions.

Having turned the out-turn data into lint weight (per seed), we may now compare lint weight and seed weight. They are evidently closely related, and the same cause which affects one also affects the other. That the relation is not absolute is shown—without the necessity of plotting correlation diagrams—by the mere existence of out-turn variations. The question therefore arises as to the causes which may disturb this relation, causing a seed to produce more or less weight of lint than is normal for its own weight.

Possible Causes of Out-turn Fluctuations.

We cannot ascribe a rise in out-turn to increased weight of individual hairs through extra thickening, for if this were so the out-turn curve should be the same as the strength curve. We cannot ascribe it to increased length of hairs of equal thickness, for this would make out-turn and length curves identical. If we ascribe it to deficient nutrition of the seed during the later stages we shall spoil our own argument, for that would entail deficient nutrition of the lint hairs, which would thicken less, and therefore weigh less. All these hypotheses and many more can be tested on the data given here, and can be found wanting; there is only one which appears to fit the case.

This last hypothesis is rather remarkable, in that it

places the cause of out-turn variations in the very last stage where one would ever think of looking for it—namely, in the open flower! At the same time it has the merit of explaining every peculiarity which ginning out-turn displays.

The Cause of Out-turn Fluctuations.

If we take the curve showing daily variations in the average lint weight per seed, and calculate the weight of lint which would be borne each day if the seeds were all of the same weight—*e.g.*, 0·1 gramme, we obtain the ginning out-turn expressed in a somewhat different form, with the seed as the foundation unit. If we now take this curve of standardized lint weight, and compare it with the curve for lint length, we find that they are closely similar when a shift of about a fortnight is made, so that conditions of the environment which are affecting the length of the lint in a fifteen-day boll are brought into line with their simultaneous action on a boll which is newly set.

There is only one way in which this effect can be accounted for—namely, by changing the number of epidermal cells which sprout into hairs (Fig. 11, *C*, *a*). To confirm this conclusion by direct observation would require the counting of all the hairs on a large number of seeds, a task which is humanly impossible by any direct method;* an indirect method which the author attempted will be mentioned below.

* While this book was in the printer's hands an article by Mr. Leake appeared in the *Journal of Genetics* (1914), dealing with ginning out-turn in the Indian cottons. By infinite patience—aided by the shorter and coarser nature of the Indian lint—he has achieved the "impossible," and shows that the differences

Looking at the significance of a high ginning out-turn in this light, its meaning is plain. If the crop of a given year has been marked by a high ginning out-turn—as compared with former years for the same variety—it implies that a great majority of the flowers have opened on days when the weather was favourable; in other words, that excessively hot, dry days, such as put a severe strain on the water-content of the plant, have been few in number. Thus the correlation which has been shown to exist between the ginning out-turn of the Egyptian crop and the size of the crop in the same year is easily understood. Moreover, if there is a sufficient proportion of good-weather days, and if there is ample water-supply, the length of the young bolls and the strength of the old ones will all be affected beneficially at the same time as the ginning out-turn of their youngest relations is being set at a high figure. High ginning out-turn is thus what it is claimed to be, an index to good quality in general.

Significance of Out-turn.

Further, if a boll has passed through severe weather in the flower stage, the immediate effect will be diminished sprouting of the lint hairs, with ultimately a low out-turn as the consequence; and in addition to this, and as a natural further sequence of it, the cells of the boll will be more or less self-poisoned, or senescent, and the later stages of development will suffer proportionately. Very

---

in out-turn between different species and varieties of them are proportional to the numbers of hairs per seed. Since this explanation has been reached by two entirely dissimilar lines of attack, we may consider it fairly well established.

bad weather at flowering, producing severe water-shortage, for example, followed by excellent weather for the rest of the life of the boll, would result in a low out-turn, rather short lint, and yet the lint might ultimately recover and thicken to normal strength.

Thus ginning out-turn is not necessarily connected with any other characteristic of the lint, except when self-poisoning is involved; but in the gamut of a series of cotton bolls ripening during a period of more than two months, the chances are that generally a high out-turn will be accompanied by generally good length, and to a rather less extent by good strength.

Special and General Behaviour of a Crop.

An interesting side-issue of this interpretation is that ginning out-turn should be more variable than lint length, and this in its turn more variable than strength, *if* we take only a uniform period of weather lasting a few days, while seed weight should be the least variable of all. Actually, the extreme percentage differences between groups of bolls ripened under the same conditions were about 2 per cent. in seed weight, 5 per cent. in lint length, 8 per cent. in breaking strain, and 16 per cent. in ginning out-turn. This is due to the different lengths of the period in which determination of the respective characters takes place, seed weight being affected over a long period, and out-turn over a very short one, so that an accidental circumstance lasting for a few hours will scarcely make any impress on the former, but will almost entirely determine the latter.

Again, however, considering the average of the chances

of a whole crop, the order of variability is reversed, short-period accidents tending to obliterate themselves; so that while ginning out-turn only changes by 1 or 2 per cent. from year to year, length may vary more, and strength so much as to mark off certain years of the Egyptian crop, just like famous or infamous vintages of port and champagne.

Before considering the grader's report on the samples from these dated flowers, it is necessary to deal with a result derived mainly from other material than the series under discussion—namely, the weights of single fibres. As in the case of breaking strain, although the measurements of this characteristic are not of direct use to the commercial growers or users of cotton as they stand, it is quite possible that some simple indirect or mechanical method of obtaining the measurements may be devised, and knowledge of them be turned to utilitarian account. The four components which could affect the weight of a lint hair are its length, the thickness of its wall, the density of the cellulose of which the wall is composed, its diameter, and its moisture-content. Length can be eliminated by cutting uniform lengths out of the middle of a fibre, and moisture by standardizing the humidity of the air in which the weighings are made, or by calculation; we do not know whether variations in the density of the wall exist, but if such is the case they could be detected by weighing hairs of equal thickness. In general, however, the weight of the hair will depend on its diameter and the thickness of its wall; thus the weights of fibres of

Weight of Single Lint Hairs.

equal diameter should be proportional to their strength as tested by breaking strain.

Lint Hair Weight Equivalent to its Strength.

The author happened to possess two pure strains of cotton whereof excellent samples were available, grown under the most suitable conditions, which had both been graded by experts as extraordinarily strong. The diameter of a hundred fibres from two such samples (strains No. 77 and No. 310) showed relatively slight differences, but the breaking strain of one was half that of the other, and the weight of equal lengths of fibre was in the same proportion. Weight determinations were made on a few samples from the series of Dated Flowers, with similar results, but, since it was not possible to construct a full series of data, it will suffice to illustrate the main points by standard samples of different cottons.

Kind.	Fibres weighed.	Weight of 10 Millimetres.	Breaking Strain.	Diameter.
		Milligramme.	Grammes.	Millimetre.
*77 G.	85	0·00176	5·74	0·0187
310 G.	77	0·00108	2·81	0·0174
*310 N.	85	0·00122	3·61	0·0176
77 D.F.	696	0·00157	4·50	—
Assili G.	362	0·00142	4·40	—

The ratio $x \dfrac{\text{fibre weight}}{\text{breaking strain}}$, where x is a constant, is almost the same in all cases—thus:

77 G.	3·26	77 D.F.	2·87
310 G.	2·60	310 N.	2·95
Assili	3·10		

Consequently, the breaking strain of a fibre is very largely determined, if not entirely, by its weight, or, in other words, by the thickness of its cell wall. This holds good between very different types of Egyptian cotton, and is independent of the lint length or of the site in which the cotton is grown. The sample marked 310 N. was grown at Neguileh, in the Northern Delta, over a hundred miles away from Giza, where the others were grown. The two samples marked with a star (77 G. and 310 N.) were respectively of Nubari and Sea Island type, and were both graded as extremely strong, or SSSS° on the grader's scale, but 310 N. was much "finer" than 77 G.

Number of Hairs per Seed.

This last result carries us on to the gradings of the dated flowers, but before leaving the subject of weight of single lint hairs it may be interesting to note that the mean lint length of 310 N., taken hair by hair, is just over 41 millimetres, and that the weight of lint on a single seed is about 0·033 gramme; since 10 millimetres of a single lint hair in its thickest part weighs 0·00122 milligramme, one hair will weigh about 0·00400 milligramme, and there must consequently be about 8,000 hairs on a single seed, whose aggregate length at 41 millimetres per hair must be 328,000 millimetres, or 328 metres.

The hairs from five seeds only of 310 N. would therefore extend for a mile if placed end to end

It is rather interesting to notice also that the sample 77 G. also works out at about 8,000 hairs per seed; for though each hair weighs more, the lint weight per seed happens to be greater.

We now turn to the results of handing the ginned lint from the dated flowers to an expert grader, in order to assess their "strength." The manner in which this was done has been described in the Appendix, and it may be taken for granted that the grader had no guide whatever as to the relation between the various samples, nór any opportunity of revising his judgments. His results can also be arranged in a curve, and smoothed to five-day means, and compared directly with the other curves.

Grading and Hair Strength.

When treated in this way it at once becomes apparent that *grader's "strength" and "breaking strain" are utterly disconnected,* and have nothing whatever to do with one another. This is a practical result of the first importance.

The most striking example of this is at the end of the curve (Fig. 21), where the breaking strain is falling rapidly to zero, but the grading remains up at SS on the scale. Comparing this with the notes given in the previous section as to the behaviour of 77 G. and 310 N., it becomes possible to see what Strength, as determined by hand-pulling, really means. The grader takes a tuft of lint to test, keeping to a uniform size of tuft; if the hairs are but slightly thickened, and consequently of low breaking strain, he takes more of them; while if they are heavily thickened, coarse, and of considerable strength individually in consequence, he includes fewer to obtain a tuft of the same size.

It is pointed out in the Appendix on p. 191 that "impact testing" of a bunch of fibres gives a result which is proportional to the number of fibres tested, and the testing which the grader applies is nearer to "impact" than to "straining." Consequently, what the grader does is to test the resistance to impact of equal weights of lint each time.

Uniformity and Strength.

But this being the case, he might be expected always to obtain the same result, and if all the fibres in a sample were alike he would do so. All the fibres are not alike, however, and so the determination of strength by the grader resolves itself very largely into a test for regularity of strength.

We can corroborate this deduction by comparing the breaking strain variability from fibre to fibre in large samples with the grader's judgment upon them. There may be a high proportion of strong fibres present, but if they are mixed with weak ones the sample is graded as "weak." This comes down finally to whether the tuft of fibres breaks under the grader's hands with a "snap" —*i.e.*, all simultaneously—or raggedly.

Having discussed the evidence obtained mainly from the abnormal conditions of a wide-sown, water-shortened, experimental plot, we will now consider the results obtained in actual field crop, since it might well be the case that the variations we induced experimentally were far greater than would ever arise in normal cultivation.

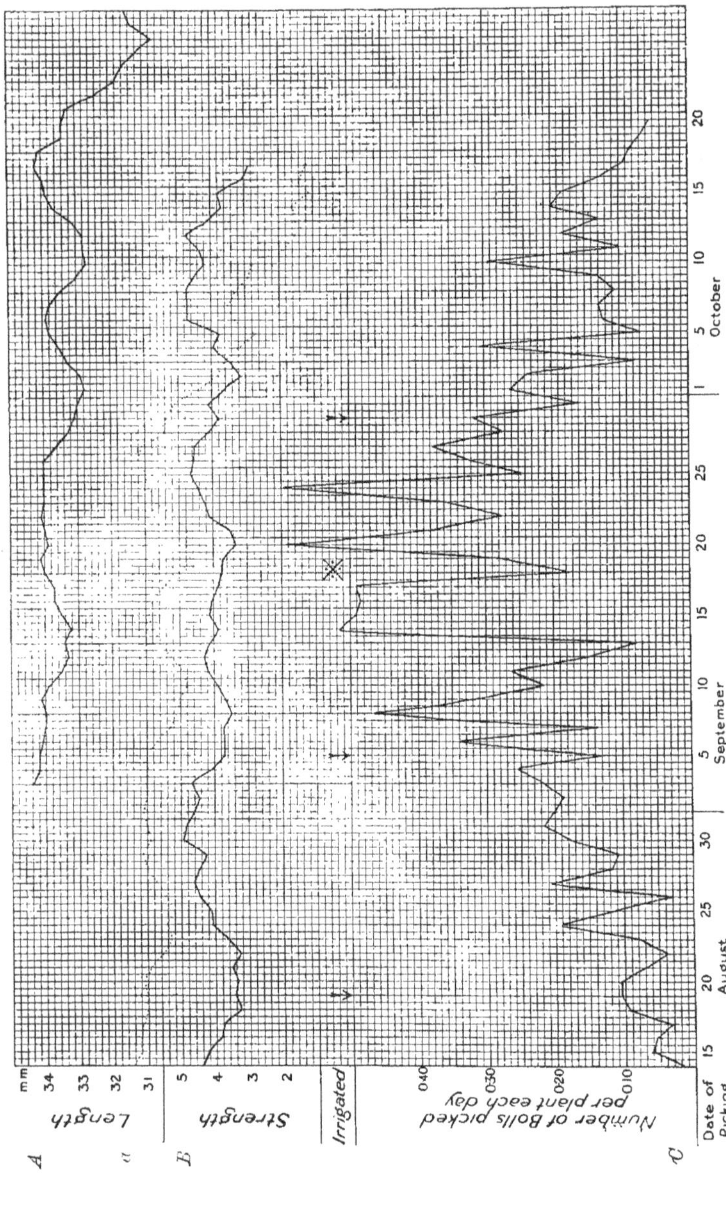
A
Length
mm
34
33
32
31
a
B
Strength
5
4
3
2
Irrigated
C
Number of Bolls picked per plant each day
0·40
0·30
0·20
0·10
Date of Picking
15
20
25
30
5
10
15
20
25
1
5
10
15
20
August
September
October

FIG. 15.

FIG. 15.—DAILY PICKINGS, 1913.

Similar to Fig. 14, p. 90, but showing the properties of the lint from *bolls picked every day* for seventy-five days off a group of plants in field crop.

From above downwards:

A, Curve showing daily change in lint-length.

a, The same curve shifted back twenty-three days, to synchronize the action of the environment upon length, with its action upon:

B, Strength of the lint hairs, which is determined when the boll is twenty-three days older.

C, Zigzag curve, showing the number of bolls ripening daily on the 130 plants employed. Note the effect of the recovery from shedding, shown on September 14, due to a watering given forty-eight days previously; also slightly shown on October 6 from the irrigation shown (arrow) on August 19.

Asterisk denotes the *beginning* of the rise of the water-table, which has the same initial effect as an irrigation.

Note how the curves of strength and shifted length begin to rise four days after each watering, and reach their maximum after eight days more. Thus, this diagram indicates the possibility of obtaining more uniform cotton by shortening the picking period, and relating *it to the dates of irrigation.*

Note also that length and hair strength *in any one boll* are entirely independent of one another.

Daily Pickings of 1913 (Fig. 15).

The series of data accumulated under this title were derived from an "observation row" of about 130 plants growing under field crop conditions at Giza, on rich deep soil, severely attacked by both ordinary boll-worm and pink boll-worm, but otherwise typical of excellent cultivation, and setting a crop of bolls which in the absence of this exceptionally bad attack of boll-worm would have weighed out at 700 pounds of lint to the acre.

The part of the field containing this group was carefully watched every day of the season, and the general health of the plants was kept as uniform as it could be, under the limitations of field cultivation. Any fluctuation shown in the cotton ripened in this experiment will, therefore, be less, if anything, than an ordinary field would show when cultivated with the same pure strain.

The strain employed was the same as in the previous series, and comparison between the two results brings out some points of interest, especially with regard to the senescence or self-poisoning induced in the previous experiment by withholding water early in the season.

The data thus obtained are directly "practical." The successive days are dated by the opening of the boll, or picking, and not by marking the flowers, while the strength was determined by "impact testing" of bunches of lint, and not by breaking single fibres.

Although the series embraces ninety consecutive days, the measurement of length on 500 seeds, and the counting of about 4,000 fibres one by one, the results can be summarized in a very few lines:—

The strength and length of each sample are again utterly disconnected. If the curve showing the length is moved backwards over 23 days it fits exactly to the curve of strength (Fig. 15).

In the water-short plot of 1912, the best fit was obtained with a 21-day shift, as against 23 days in this experiment, indicating that the senescent condition of the 1912 plot shortened the maturation period by checking growth. It is common knowledge that water shortage "ripens off the crop"; we can now see the price which is paid for it; namely, weakening and shortening of the lint, and reduction in actual yield. The crop which is thus being hustled into maturity will look better than one which is being allowed to take three or four days longer over its duties, because the withering and falling of the leaves exposes the open bolls to view, but a count of the actual numbers will show that there are not so many.

Hastening Maturity.

Comparing the lint length in this new series with that in the old one, we find that, whereas the plot which had been starved of its due water allowance only once exceeded 31 millimetres on the five-day means, this properly-cultivated field-crop series only drops below 33 millimetres five times before October 22, and does not touch even the maximum of 1912 till it has finished cropping. The maximum reached in this series was 34·4 millimetres instead of 31·1 (*cf.* Figs. 14 and 15).

Length of Lint.

The next point of interest in regard to lint length is the

fluctuation it shows, even under conditions which were almost ideal for a field crop. Obvious senescence appears to have been excluded entirely, but in spite of this the length swings steadily over a range of 1·4 millimetres on the five-day means, which in actual daily data of the same precision would be nearly 2·5 millimetres, or $\frac{1}{10}$ inch. The length of the lint as shown in these experiments is determined by combing on the seed; if expressed as the length of a "pulled tuft," in the usual way, this fluctuation would be stated as a change from about $1\frac{5}{8}$ to $1\frac{3}{4}$ inches, which, though not great, is still appreciable in the manufacture of combed yarns.

Fluctuating Length.

Lastly, it will be seen that after the lint length curve has been shifted 23 days to fit the strength curve, each irrigation shows an effect upon it. On the third or fourth day after watering the two superposed curves begin to rise, continue to rise until about the tenth day after watering, and then die away again, to be revived by the next irrigation.

Effect of Irrigation.

Adding on the 23 days of the shift, this means that the length shows the first signs of having been affected by watering in those bolls which open 27 or 28 days afterwards, and that the maximum effect of watering upon the length is shown in those bolls which open about 32 days afterwards.

To obtain the age of the boll at which the effects of watering are produced, we must subtract these intervals from the maturation period; this period was determined for this series by an indirect method as being 48 days.

Therefore the maximum effect on lint length is obtained when the boll is 15 or 16 days (48 – 32 = 16) old at the time of watering, while if the boll is more than 21 days old (48 – 27 = 21) it is unaffected.

This result is different in degree from that which we obtained in the previous series after the long water starvation, and it indicates that in this case we are dealing with a simple and straightforward limiting factor effect, instead of the complicated poisoning effect. In this case the effect of watering is simply to allow the lint hairs to grow more during the last few days than they would otherwise have done. The general principle, that the maximum effect is produced round about the fifteenth day of boll development, is the same as before.

Turning now to the strength curve, we may first note that it is immaterial for general purposes whether the strength is determined by breaking single hairs, or whether bunches are tested by impact. The impact figures obtained in this series cannot be converted directly into breaking strains for direct comparison with the previous series, but some determinations of the breaking strain made tediously by hand—in the absence of the author's automatic tester (Pl. XVI.)—indicate that the maximum strength attained in this series was rather higher than the maximum attained in the former series, just as in the case of lint length. Moreover, whereas the former maximum was rather spasmodic, the strength in this series under good field conditions is maintained steadily for several days at a time.

Strength of Lint.

If we next consider the extent of the change which the strength undergoes between the effects of one watering and that of the next, we find that it is very much greater than in the case of the length. The change which we found between 32·7 and 34·4 millimetres in the five-day means of lint Length stands in the ratio of 95 : 100. In the case of lint Strength the extreme values recorded stand as 65 : 100. This greater capability for fluctuating is not due to accidents of method, for the fact that our impact test determinations are less precise than those of length should help to obliterate such distinctions. We may safely assert that in this series the strength of the lint rose from 60 to 100 in a single week (August 22 to 29).

Greater Fluctuation in Strength.

This phenomenon introduces a proposition of very practical interest to the spinners of fine cotton. Would it not be worth while to encourage the practice of picking at shorter intervals? If we regard these curves as an analysis of the three "pickings" in which the cotton crop is usually harvested, it becomes clear that the composition of the lint at any one picking cannot be uniform, except by accident against long odds. In conventional practice the bolls which we have studied here would have been gathered in three groups or pickings, on or about September 10, September 30, and October 20. By reference to the curves it will be seen that pickings taken on any of these three dates would have been irregular in length and in strength, since they would consist of all the bolls ripened before that date. In general experience the first

Short-Period Pickings.

picking is the best, and the third the worst. The differences are partly due to increased frequency of insect attack, and partly to the fact that the third picking, if delayed too long, will include senescent lint, especially if cultivation has aimed at hastening maturity by deprivation of water, or if the water-table has risen in the meanwhile.

Quality of Pickings.

As in the case of other common experiences which we have discussed, this difference between the pickings is a matter of accident—excluding boll-worm—and not of direct and necessary causation. The temperature in Egypt during all the period from July to mid-October is rarely a limiting factor directly, so that it is not till late October that the cultivation of cotton in middle Egypt becomes dependent on the temperature; if the cultivator will abstain from trying to save a day or two in maturity by cramping the water-supply (and it should be noticed that the Fellah himself never does so unless he is obliged), and is lucky with the boll-worm attack, there is no reason why he might not obtain cotton of the same value in his third picking as in his first picking, excepting that the autumn fogs may cause the lint to mildew if left too long on the plant after the boll opens. The whole thing is a question of accidents; there are more chances of unfavourable accidents late in the season than in the beginning of the season, but most of them can be avoided. The worst of all these accidents in Egypt is the rise of the water-table, which takes place in permeable soils when the flood comes down.

In the year 1913 the Nile flood was later than it had been for over a century, and lower also. There is therefore very little effect visible in these curves, as compared with what a normal year would have shown, excepting a beneficial one due to capillary damping of the supernatant soil on and after September 15 (star in Fig. 15), when the effect of the previous watering was dying away; since the water-table in this particular plot of the Daily Pickings did not rise nearer to the surface than 1·30 metres, no very striking effect could be expected so late in the year. The 1912 results from daily flowers were also unsuitable for demonstration of the water-table effect, since wide-sown cotton does not suffer nearly so much from this cause as do the closely crowded plants of field crop. Nevertheless we can draw our own deductions from existing evidence about the effects of the water-table on other growth-processes in a field crop of cotton, and from what we have already learned about the development of the lint, with the following conclusions:

Water-Table and Lint Quality.

When the water-table rises so as to immerse the lower half of the root system of a field crop of cotton in Egypt, the effects will show up in the following order: Bolls opening ten days later will have weak but long lint, those opening five weeks later will have lint both weak and short, with a high ginning out-turn, and those opening seven weeks later will be worthless in all respects.

Returning now to our original statements as to the possibility of growing equally good lint at any time during the season until the falling autumn temperature becomes a limiting factor, and barring accidents such as boll-worms and water-tables, let us see where the longest

Differences between Pickings.

and strongest lint was produced in this Daily Picking Series. The lengths of the earliest bolls have not been determined, but their weakness puts most of them out of the running; those of September 3 were good in both respects, but only slightly superior to those of September 26, and on October 9 we have another period which is much the same. An important conclusion follows from this search for good samples: it should have been possible so to adjust the irrigation intervals that the length and strength of the samples, although determined at 23-day intervals, should move together. The obvious way of doing this is to keep the intervals between irrigations constantly related to 23 days for this strain, and to 20-26 days for other strains and sites.

It may not be entirely coincidence or convenience, as is commonly assumed, that the rotation of water in the

Canal Rotations.

Egyptian canals is commonly arranged nearly at this interval; the usual explanation is that the cotton cannot stand longer intervals without water, but this is not strictly correct. Personally the author considers that lighter waterings at 11 and 12 day intervals alternately would be better still, and would of course produce the same result.

Leaving this point, and assuming that the irrigation intervals had been so adjusted as to change length and strength simultaneously, this would not abolish the net change of both. Thus the commercial pickings would still be composed of samples showing various properties. If, on the other hand, the intervals between the pickings were shortened, and the pickings kept separate, much greater uniformity could be attained—it might even be permissible to mix certain pickings together; thus, in the series here discussed, a much more level sample is produced by mixing September 1, 2, 3, 25, 26, 27, and October 9, than by picking all the bolls which ripened between August 22 and 29. Such mixing would not be permissible for the unskilled cultivator, but there are some clues upon which he might guide his conduct.

Classification of Pickings.

For example, if strain No. 77 were being cultivated at Giza, and if short-period picking were practicable, and if the irrigation intervals had been adjusted to bring strength and length fluctuations into step, then the grower would keep all the bolls which opened between the sixth and twelfth days after each watering separately, and either mixing them or keeping them separate, according to opinion obtained, would dispose of them as the highest grade of his product.

Uniformity in Cotton.

Discussion of this matter brings us to the ultimate analysis of "regularity." If pure strains are cultivated, and if pickings are so taken as to include only those days on which the cotton is all alike from day to day, there remain only two more ways in which the sample can be made irregular. These two are, firstly, individual fluctuation from boll to boll and from plant to plant, and, secondly, variation in lint properties of various parts of each seed involved.

The first cannot be eliminated in field crop conditions. If the plants were separated by intervals so wide that each one had all the soil it could occupy, if such spacing did not bring in secondary difficulties, and if the soil were absolutely uniform to a depth of 3 metres, then there would be no fluctuation from plant to plant, but there might still be differences from boll to boll. Conversely, there is a strenuous struggle for existence between plants crowded in field crop. This struggle is subterranean and invisible, but none the less bitter. Under Egyptian conditions, the success of any individual over its neighbours is partly rectified above-ground, since its consequently greater growth brings shade to them and lessens the strain on their root systems; but if a pure-strain population is closely crowded in a shallow soil, it will be found by the end of the season that a few plants have alone survived to mature their seed, the remainder being wizened sticks.

Plant-to-Plant Fluctuation.

The last component of "regularity" is the distribution over the seed. This can be partly controlled by the use of suitable strains, since the character of the distribution is inherited, and strains can consequently be isolated in which—when properly grown—the mean length and strength of the lint varies but little from the tip to the butt of the seed. In periods of unsuitable nurture, however, even these strains will make poor lint at the tip of the seed, farthest away from the food-distributing centre, which is situated where the incoming vascular bundles ramify in the butt-end of the seed-coat. All the changes we described in

Hair-to-Hair Fluctuation.

the previous chapter, such as the development of the palisades in the seed-coat, and the lengthening and thickening of the lint, appear to be begun simultaneously all over the seed, but completed first at its butt or thick end. Therefore, if the development of any feature is checked a little too soon, irregularity will follow.

A natural consequence of this is that periods of change are also periods of irregularity. This applies not only to distribution on the seed, but also to the mean maximum length from seed to seed and boll to boll, as may be seen by examining the figures in the Daily Picking Series. When the conditions of the environment are kept constant for any length of time, the lint ultimately ripening from day to day may be good or it may be bad, but it will at least be uniformly one thing or the other, and as such it will be saleable for a definite purpose. On the other hand, when the conditions are changing from one day to another as when soil is drying up—the slight variations in the length of the maturation period from boll to boll will in themselves be sufficient to mix better bolls with worse ones.

Uniformity and Changes of Environment.

Variation from seed to seed within the boll has been extensively discussed by other writers, but its case is covered by the preceding account.

Summarizing the results of this chapter, we have a result of fundamental importance in the confirmation which these Dated Series give to the deduction drawn from developmental evidence, that the same conditions of the environment affect the length and strength in the same way, but in bolls of different ages. The interval, in the case of a certain pure strain in a certain site and season, was twenty-three days.

Summary.

Absolute regularity of lint is unattainable, even with a pure strain, but apparently much might be done to improve in this direction by shortening the picking interval with discretion. Whether the cost of so doing would exceed the cost of producing the same result by Combing remains to be seen.

CHAPTER V

THE DEVELOPMENT OF COMMERCIAL LINT

In the previous chapter we have seen how some of the more obvious properties of the seed-cotton and lint are attained, and we have found that the story is an extremely simple one. If we avoid technical details, the whole matter resolves itself into this: that each characteristic depends on the reaction which takes place between the constitution of the plant and the circumstances of the environment, at the time when the characteristic is being built up by growth.

The minor circumstances which have tended to obscure this main issue are somewhat as follows:

(*a*) The climatic circumstances under which cotton grows are not favourable to the sustained and detailed research required.

(*b*) The age of the boll cannot be dated entirely by its external appearance.

(*c*) The results of self-poisoning or senescence have not been separated from the simpler direct effects of the environment.

(*d*) Until the conception of Limiting Factors was introduced by Mr. Blackman, the analysis of environmental effects was impossible.

(*e*) Until pure strains of cotton were available, the differences from plant to plant were sufficient to obscure the differences from day to day.

Having analyzed the problem of development down to its simpler constituents, it remains to build back into complexity. Interesting though our analysis may be, in itself it shows that the commercial crops are not likely to approximate to any sort of ideality for many years to come. We must therefore attempt to sketch the relation of these facts to the cops and hanks of yarn which the mills produce.

Spinning and Growing.

This task is one which any author might well forego with pleasure. If his acquaintance is with the grower's side of the matter—as in the present case—he cannot have more than an inkling of the work and problems of the mill; while if his knowledge is sound on the spinning side, he is not likely to have spent sufficient time in the cotton-fields to be familiar with the limitations which debar him from obtaining the ideal cotton.

There can be no doubt that the wisest course for the present author would be to leave the raw cotton in the open boll, but to do so would defeat one of the motives which have led to the writing of this book—namely, a desire to establish a common language between grower and consumer. A great deal has been done in this way of late years, especially as regards Egypt and Lancashire, and some of the most outstanding misconceptions have been abolished; but all that which has been done is not a tithe of what remains to do. When the Spinner can prescribe the cotton he desires in such a way that the Grower can read the prescription, and can in his turn set to work to sow the seed and grow the plants in such a way as to obtain it, the cotton trade will be homogeneous, and capable of a high degree of efficiency. That the trade of spinning alone is enormous, specialized, and intricate, comprising many separate trades within itself, will be admitted by everyone. It is probably not quite so intricate an organization as a cotton-plant, and its components have the advantage of articulate speech and of historical origin. So long as the Grower of cotton was content merely to do certain things because past experience had shown that they were, on the average, the best things to do, he could not expect much sympathy when he complained that the language of the Spinner was not comprehensible to him. At the present time, even though there is no millennium in sight, and although all the knowledge we are acquiring may not pay anyone to apply, there is a definite tendency towards this establishment of communications between the two ends of the cotton trade. The forces of curiosity are getting out of hand. Each end is beginning to wish to know more about the other end's business, and to realize that the Cotton Trade is not confined within the walls of the mills.

Now, when such intercommunication begins on the feature, let us say, of "strength," the common language is at once conspicuous by its absence. Well-intentioned authors write at length on the strength of single fibres tested by straining and by blows from a falling weight, graders take samples of lint and

"Strength."

state their opinions as to strength with uncanny accuracy, and the spinner expresses most varied opinions as to the strength of the yarn he produces from that lint, according to the class of yarn he is making. The question naturally arises as to what "strength" is, and each person concerned gives an entirely different answer, which is quite correct in every case.

It should not be beyond human ability at least to construct a series of analyses in the precise form which science exacts, so that, even if the gap between grower and spinner cannot be bridged at one jump and by one man, a bridge might at least be built. It is with the intention of starting one abutment from the grower's side that this essentially impossible chapter is included in these pages, and the author hopes that criticism by the spinner may be tempered accordingly

It will save endless reservations if we first deal with the lumber brought in by varietal impurity, and clear it out of our way. There are no pure cottons in commercial cultivation at the present day. This statement is necessarily based on negation, but the standard varieties tested by the author for their composition are some fifty in number, including Sea Island, Upland, and Indian, as well as every known or unfamiliar Egyptian variety, and several semi-wild cottons. From any of these a number of strains can be isolated and bred true, or pure forms can be "split" out of individual hybrid plants.

Impurity of Commercial Varieties.

It is possible, though very improbable, that a variety

might consist of a dozen different strains, and yet all the plants might produce exactly the same lint. Perfectly definite differences in petal colour and suchlike can be quite independent of lint properties. Similarly, so far as our knowledge goes, white and brown lints may be otherwise identical, or two entirely different lints be borne on plant bodies which are externally indistinguishable. There is no limit yet known to the shuffling of characters which may take place in this way through deliberate or natural intercrossing within the main groups of the genus.

Independence of Inherited Characters.

The obvious characters make only half the story, however. We are beginning to realize, and in some ways to understand, how two kinds of cotton may show entirely different reactions to their environment, so that such a complication as the following example presents is well within probability: Two strains of cotton are growing a hundred miles apart, one on the sea-coast, the other in the interior, and appear to be exactly the same. When grown side by side in either site, the imported one is conspicuously unhappy, and plots of it can be recognized at the other side of the field. These differentiating characters which are not obvious include such reactions as tolerance of salt, liability to shedding, stage at which senescence sets in, and velocity of growth at given temperatures. In separate species of the genus these features may be most clearly seen, but in a less obvious degree they may be found in the strains isolated from the same variety. One of the most interesting sights of cotton-growing in the

Growth Habits.

PLATE IX.—CONSTITUTION OF COMMERCIAL VARIETIES.

Photographs of three representative plants in families descended from three plants of the Abbassi variety. One-thirtieth natural size, taken just before the opening of the first bolls. The three parent plants were chosen first by their seed-cotton, as lying in the centre of the Abbassi Target diagram (Fig. 16, p. 134). The lint was then submitted to an expert and graded as Abbassi. The three photographs only convey a faint impression of the differences between these three sets of offspring. Left: Wiry stem, semi-prostrate habit, practically no vegetative branches, leaves very small, joints long. Centre: Robust erect stem, slight development of vegetative branches. Right: Stout stem, semi-prostrate habit, vegetative branches strongly developed (*e.g.*, main stem is the left-hand shoot in the photograph).

author's experience was the annual growth-race between strains of cottons derived respectively from Willet's Red-leaf, King, Asia Minor, and the average Egyptian plants; first one and then another of the competitors would lead on the same dates each year, according to their specific peculiarities.

Natural Selection.

If the differences between the components which go to make up a commercial variety were confined to mere structure and colour, there would be very little material for natural selection to lay hands upon; but since there are also these physiological differences, it follows that some kinds of plants flourish best in one locality, and produce more seed, with the result that the sowing of the next season contains more of these plants, and the general properties of the variety alter accordingly. The name given to this alteration varies: if the change does not spoil the lint, it is called "acclimatization"; if the lint of these flourishing plants is inferior, the change is called "deterioration."

Uniformity and Cultivation.

Now, a commercial variety of cotton growing in any one site and year is made up of many different strains of cotton and of hybrids between them. Some of these plants are well suited to their environment, others are not. If cultivation is very good, or, in other words, if as little tax as possible is put upon the self-regulating functions of the plant, even those plants which are comparatively unsuited to the environment will grow fairly well, and will make as good lint as they can; the crop resulting will therefore be as uniform as it can. If, on the other hand, cultivation is poor, only those plants which happen to be thoroughly well suited to the environment will produce tolerable lint, and the presence of this lint in the sample will increase the irregularity of the sample.

Thus, if a commercial variety is well cultivated, it tends to greater regularity; and if badly cultivated, to greater irregularity. Since the brightness, cleanliness, lustre, etc., of the lint are all indices of good cultivation, they also go hand in hand with an approach to regularity.

On the other hand, no amount of good cultivation can make a short-staple plant into a long-staple one, so far as our present knowledge of growth—and especially of senescence—can avail us. Consequently, any approach to real uniformity is impossible with impure varieties. It may possibly be thought that undue insistence is here being laid on varietal impurity; that the persons who introduce new cottons are not likely to introduce them in an obviously mixed condition; and that in speaking of such impurity the author is applying some hyper-critical botanical test.

The Limitations of Good Cultivation.

There is a very simple way of presenting data for varietal composition in respect of two commercial characters simultaneously, based on the statisticians' Correlation Diagrams, which we may term Target Diagrams, since the scatter of dots over the diagram is used in the same way as the shot-pattern of a shot-gun. We have seen that all these measurable characters, such as lint length and ginning out-turn, fluctuate to a definite degree round a mean value, within a pure strain. If, therefore, we grow a family of pure-strain plants, under uniform treatment, and determine the out-turn and length for each plant in a certain period of the season, we can draw curves showing the distribution of the variations in each respect through the family. If we now place these curves on two adjacent sides of a square, we can make a target diagram in the following way: Find the position of plant No. 1 in the length curve, and draw a line into the square at right angles to the side along which the length curve is plotted. Then repeat the process with the out-turn curve. At the point of intersection of these two lines make a conspicuous dot. Repeat the process for each plant, when it will be found that the group of dots thus made will give a picture of the amount of "scatter" in both characters at once, instead of only showing one at a time as the curves did (Fig. 16, p. 134).

If we are handling a pure strain in this way, the scatter in either character will be definite and regular, and the dots will form (with suitable plotting) a circular group, the centre of which is densely dotted, while the dots become fewer and fewer towards the margin of the group (Fig. 16, Targets 5-7, 10). The centre of the group lies at the mean for each character.

If we now mix two strains together, which are distinguished in their average out-turns and lengths, the dots will form two groups. If the two strains are very widely different, the groups will be quite separate. If they are only slightly different, the fact will still be recognizable when the diagram is viewed from a distance, for instead of blurring into a circular arrangement, with the darkest spot in the centre, the blurred diagram will be more or less oval, and instead of having one darkest spot in the centre, it will show two spots.

Again, if we take a pure strain which is suspected to have been contaminated by crossing, and make the diagram for all the plants, we shall probably be able to detect the hybrid rogues, since it is probable that their out-turn and length will be unlike that of the pure strain, and consequently the dots representing them will be likely to lie well outside the group. If we have four such measurable characters we can plot six such diagrams, and be practically certain of finding the rogue in one or more of the six.

With this description we may further examine the target diagrams (Fig. 16) which show the composition of the principal commercial varieties of Egyptian cotton, not in respect of any abstruse botanical features, but in the directly commercial characteristics of lint length and ginning out-turn. Side by side with them are plotted (Targets 5, 6, 7, 10) the target diagrams for pure strains, to compare the amount of scatter which need exist with that which actually does exist. The objection may be raised that the various plots might have received different treatment, hence accounting for the different degrees of scatter, but this is not the case; when the plants are grown for such comparison, they are all mingled together on the same piece of land, plant by plant, so that all share equally in any variation of soil or of cultivation.

Impurity of Egyptian Varieties.

One or two points in these diagrams, though of particular interest to Egyptian cotton-growers, have also some general significance. It will be noticed that Ashmouni (Target 9), the Upper Egypt variety, has a fairly compact diagram; this is mainly due to the fact that it has been comparatively isolated in Upper Egypt, and has not been offered the same opportunities for admixture as the Delta varieties ; at the same time it is by no means pure, and in other respects—such as lint colour—shows itself to be quite heterogeneous. The Delta variety of Egyptian which has deteriorated least, according to the spinners, is Yannovitch (Target 11), and the compactness of its diagram confirms this. On the other hand, Afifi (Target 12) has deteriorated very badly (according to the same authorities), and its target diagram shows a very wild scatter. Only one other variety is as wild, namely, Assili (Target 8), which was only introduced in 1910, showing that it was not so pure as it was at first claimed to be. In the case of this variety we can trace the deterioration by target diagrams for the best commercial seed of successive years (Targets 4 and 8), and can watch the gradual obliteration of the "type group" of dots by mixture and crossing with the outlying rogue dots. Sakellaridis is another new variety, also not as pure as it ought to be.

Relative Importance of Seed and Cultivation.

These diagrams have been included to show that varietal impurity is a real difficulty, and not merely the theoretical matter which it might appear to be. If a variety of cotton consists of plants which, when growing side by side, produce lints differing in length by as much as half an inch, it is waste of time to advise short-period pickings

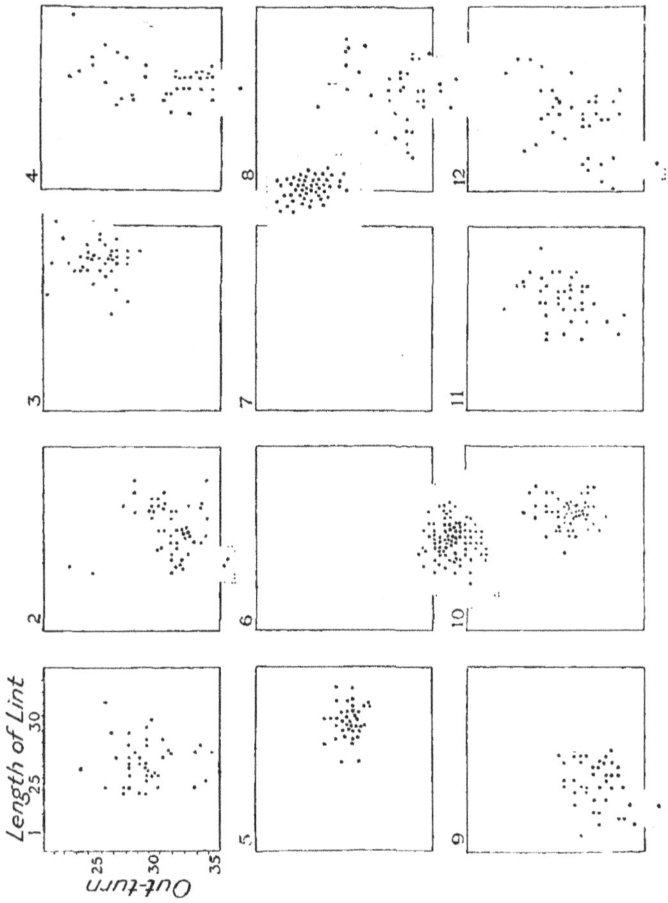

Fig. 16.

FIG. 16.—TARGET DIAGRAMS SHOWING THE COMPOSITION OF VARIETIES AND PURE STRAINS.

Each of the twelve targets is a square, bounded as follows: Above, by a ginning out-turn of 21 per cent., and below by one of 35 per cent.; to the left by a lint length (combed) of 20 mm., and to the right by one of 34 mm.

Each dot represents a single plant.

The position of each dot thus shows the lint length and the ginning out-turn of the plant it represents.

If all the plants are identical in these and other respects, all the dots come together, except for accidental displacement. If the plants are not alike, the dots must obviously be scattered apart.

To realize the use of the diagram clearly, it should be viewed from about twelve feet away. Only the targets 5, 6, 7, and 10 are then clearly visible; 9 and 11 are faintly so. The others have their dots so widely scattered as to be invisible.

The kinds of cotton represented are—

1. Abbassi	2. Nubari.	3. Sakellaridis.	4. Assili of 1911.
5. No. 77.	6. No. 111.	7. No. 310.	8. Assili of 1912.
9. Ashmouni.	10. D. F.	11. Yannovitch.	12. Afifi.

"D. F." represents the daily samples of strain No. 77, discussed on pp. 90-109.

for the sake of reducing that irregularity which is due to environment, and which does not amount to more than a quarter of an inch (see Chapter IV.).

As in the case of other peculiarities of the crop which we have discussed, the lint properties are therefore average properties, the deviations from the average being widest in bad cultivation, and least when cultivation is good. To conclude with an instance which is perhaps more striking, and certainly simpler, than these target diagrams. The colour of a sample of ginned cotton is an average colour. Thus the modern Ashmouni cotton has deteriorated in colour, losing the full, rich golden-brown which formerly characterized it, and becoming much paler. If we take a prize sample of modern Ashmouni, and raise thirty or forty plants from it, we shall find a few plants whose lint is nearly white, many creams and light browns, and perhaps about one quarter of the plants will be found to bear rich brown lint, which, when placed side by side with a sample of twenty years ago, matches exactly. The colour of a prize sample of Ashmouni is thus produced by placing together hairs of these various colours, the lighter hairs diluting the colour of the darker ones. The old colour can immediately be restored by propagating some of the plants which bear the rich brown lint.

Colour Impurity.

The photographs we have given (Pl. IX.) show very imperfectly the differences from plant to plant within a variety. Cotton is a difficult photographic subject, except when single plants are taken with a background, and in that case much of the impression is lost, as compared with the view of a series of plots. The examples given (Abbassi) are interesting; in the first place, the target diagram of Abbassi (Fig. 16, Target 1) was made, and then nine plants were chosen from the type portion of the target group; all were graded and pronounced to be of Abbassi type, besides having the Abbassi colour. The seed from these nine plants was sown in nine adjacent plots, and in seven out of the nine the strains were pure as far as branching and leaf-form characters were concerned. None of these seven kinds in the least resembled one another, and all were more dissimilar than the plot of Abbassi had been from other Egyptian varieties in the previous year. From those plots with peculiarities which showed up well in the camera, photographs were made of average plants; and, striking though these differences appear in the photographs, they decidedly minimize, rather than exaggerate, the differences actually shown.

Plants of One Variety.

The commercial sample of lint thus consists of different kinds of lint mixed together, these different kinds having been borne on different types of plant, with different methods of reaction to their environment. Over and above all this are superposed the effects of the environment from day to day, causing variations in the length and breaking strain of the fibre in successive bolls, as we have shown, while incidentally there are differences between the individual experiences of particular plants of the same kind, and differences from one site to another. It follows from

Origin of the Commercial Product.

this complexity that, while an amateur can grade the lint from small plots of a pure strain fairly successfully, it takes an expert to grade commercial lint.

Properties of the Commercial Product.

Our next step is to attempt to connect these properties of the lint as shown by our methods, with those recognized by the grader, with those ascertained finally by the spinner, and to see how far our analysis will carry us.

Leaving the major characters of length and strength for the moment, there are minor characters such as colour, cleanliness, elasticity, and lustre. Elasticity is probably involved in uniformity; all cotton fibres are elastic to a high degree, and the resulting "feel" of the cotton in this respect is probably a combination of effects resulting from uniformity, fineness, and twist.

Colour Modifications.

Colour is, in the first instance, based on the inherited peculiarities of the variety, modified by subsequent events; if the seed-cotton remains too long on the plant, or is exposed to dew or strong sun, the colour fades; there also appear to be different inherited degrees of "fastness" of the colour, some strains bleaching more easily than others under similar conditions of exposure; consequently, while it is often contended that cultivation in a new country changes the colour, and while such change is quite probable, since colour fluctuates like any other character, many such examples can be analyzed to fading-phenomena when they are shown in the first year, or to natural selection in a mixture (such as Ashmouni mentioned above) when shown in subsequent years. Cleanliness hardly needs mention from our point of view, being the result of accidental episodes which happen after the boll has opened, or of insect attacks on the boll.

Lustre is somewhat peculiar, It would seem to have been insufficiently analyzed by investigators, and in the first place it is probably almost synonymous with twist; if all the hairs in a sample are well and evenly twisted, there will be an infinite number of convex surfaces, each reflecting a spot of light. In addition to this there is refraction of light, which may be seen by holding a well-twisted fibre against a dull background with a good north light well overhead. The fibre then shows slight diffraction colourings; this undoubtedly has considerable influence on the appearance of the sample, through the reflection of light back from the concave surfaces inside the fibre; it would necessarily involve the translucency of the fibre wall, since any opacity of the wall would obliterate it. Such opacity might result from prolonged exposure of the seed-cotton on the plants, or from irregular deposition of the thickening layers of the wall, in so far as single fibres are concerned, while irregularities from fibre to fibre would have a similar effect in a bulk sample of lint. In former discussions of lustre the major importance has been attributed to the cuticular skin of the fibre, and to variations in its reflection of light; but cuticle is one of the last plant tissues to be affected by ill-treatment, and it seems more than probable that the causes of lustre changes and variations lie behind the cuticle.

Lustre.

One point may here be mentioned which the author has

been unable to elucidate, namely, the habit in the U.S.A. of storing seed-cotton for a month if possible before ginning,* and the reverse habit in Egypt of ginning as soon as possible after picking.

The Storage of Seed-Cotton.

This habit in the U.S.A. has always been ascribed to the protoplasm of the lint hair cell remaining alive, and, so to speak, finishing its work on thickening the wall while in the store. Upland cotton does not behave in this way when grown in Egypt, the cell-contents dying as soon as the boll opens; and either the accepted explanation is incorrect, or else the process is merely one of "conditioning," by effecting a more uniform distribution of moisture through the sample than when it was first picked. Sometimes the effect is ascribed to the oil from the seed working its way into the lint, but the author is not aware that any chemical proof of this statement has ever been brought forward, nor is it easy to see how oil (which is buried as droplets in the living protoplasm of the embryo only) can work its way out, through the dead tissues of the seed-coat, including the vegetable ivory of the palisade layer (Fig. 20), and ultimately into the lint.

We now turn to the major characters of length and strength, with the all-embracing feature of uniformity.

There is very little to be said regarding Length, other than our previous remarks in the foregoing chapters. If any of the many contributing causes have brought about irregularity in this respect, the sample will be unsatisfactory; it will not "pull to a hard edge" (Fig. 18, *B*),

* But not in Texas.

and, since even the best machinery cannot completely equalize the distribution of tufts of varying length in

Length.

the yarn, it will make weaker yarn than a similar sample in which the lengths of all hairs are the same and equal to the mean length of the irregular sample ; the old adage that the strength of the chain is that of its weakest link applies very completely to yarn strength.

The discussion of Strength embodies this adage, and also very much more. In the first place we must dis-

The Kinds of Strength.

tinguish between the different kinds of "strength"; there are primarily, breaking strain or hair strength, hand-impact-testing or grader's strength, and the strength of the spun yarn. Secondarily there is the strength-variation from fibre to fibre, which affects all three classes.

Hair Strength.—We have seen good reason for believing that the breaking strain or impact resistance of single fibres is almost entirely dependent on the sectional area of their walls, independently of their diameter.

Worked out in this way, it is not without interest to note that the tensile strength of a cotton fibre is about double that of wrought iron, so that lint is not quite such a delicate and fragile substance as one is inclined to imagine it. If the wall thickness is the same in two hairs, the hair strength will be proportionate to the diameter of the hair cell; while if the hair cell diameter is the same in both, the hair strength will be proportionate to the thickness of the wall. Our only reservation in this respect is that it is possible for the texture of the wall to affect the strength somewhat ; but—just as flowering is the main determinant of yield—wall thickness must be the main determinant of hair strength, while texture can only affect it secondarily as to strength, though possibly greatly in regard to lustre. We have also seen that hair strength is a definitely inherited peculiarity, though subject to much greater fluctuation than length, some strains never making a very thick wall, while others make very thick ones whenever they are given an opportunity.

A side-issue of great importance from this discussion of hair strength relates to the diameter of the fibre.

Diameter.

This we have seen to be comparatively constant within a pure strain; but since there is a little uncertainty in some previous writings on this subject, it may be well to discuss the matter more fully. The diameter of the developing hair cell is fully attained almost immediately, though the position of the maximum diameter in the full-grown lint hair may vary with different kinds of cotton, and cannot be settled till the hair has grown to its full length.

This diameter is that of a very thin-walled and more or less cylindrical tube. The diameter of the ripe lint hair

Diameter affected by Hair Strength.

is quite different: the walls of the tube have been thickened, and then have collapsed (Fig. 11). Obviously, if the thickening has been negligible, the width of the ribbon thus formed will be one-half the circumference of the original tube. If, on the other hand, thickening has been so complete that the tube becomes a solid rod, it will not only be incapable of collapse, but its diameter will be the same as that of the tube. The diameter of the ripe fibre may thus vary from 157 to 100, according to the amount of thickening, where 100 represents the diameter of the original hair cell (Fig. 17). Thus, in terms of width of the ribbon, the more a hair cell is thickened, the finer it will be, which is obviously absurd; thus the term "fine-

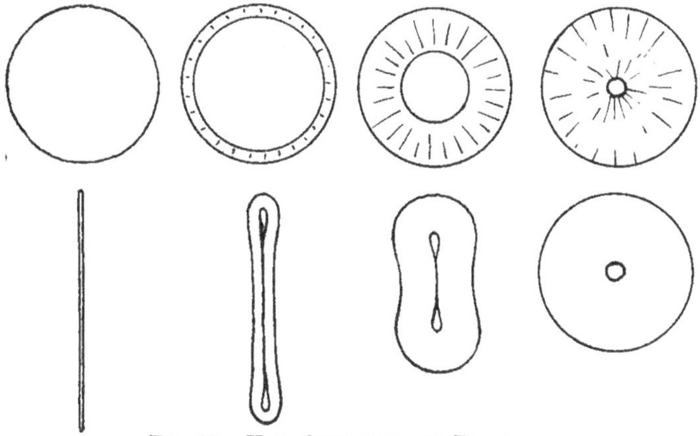

FIG. 17.—HAIR STRENGTH AND DIAMETER.

Diagrammatic transverse sections of lint-hairs, indicating how increased wall-thickness—*i.e.*, hair strength—must lessen the maximum diameter of the ripe hair.

Left, no thickening. Right, excessively thickened. Above, before the boll opens. Below, ripe and collapsed.

ness" does not relate to the width of the ribbon, and has little to do with the maximum diameter of the collapsed cell.

If again we compare the two pure strains already mentioned (p. 106), which had practically identical diameters, but very different hair strengths and very different designations in respect to fineness, we see that the maximum diameter of the lint has nothing to do necessarily with comparisons of fineness from one kind of cotton to another, as is so frequently stated.

Fineness.

On looking through published figures on this subject, one is struck with the way in which data for diameter have been stretched to make them fit the view that diameter and fineness are equivalents. The extreme range of mean fibre diameter in good samples of all commercial cottons may be taken as 0·016 millimetre for Sea Island, and 0·025 millimetre for some Indian cottons. The squares of these numbers stand in the ratio of, roughly, 2 : 5, and variations of this magnitude in fineness may be found within Egyptian cotton alone, where the diameters are practically constant. If, however, we consider the thickness of the wall of the fibre, which cannot well be measured except by cutting sections, and so obtain figures showing the thickness of the walls, they will follow the gradings for fineness much more closely.

Thus we reach a definition of fineness as practically equivalent to hair strength. Fineness is partly a matter of cell diameter, but more a matter of wall thickness.

The statement just made, that a fine lint is weak lint, may at first seem to be a *reductio ad absurdum*, but it needs further consideration from the grader's point of view.

Grader's Strength.—In the preceding chapter we saw that grader's strength and breaking strain were entirely independent, and that not merely from day to day in the same kind of cotton, but also between different cottons. We also concluded that the grader tested the lint for its resistance to impact, and that his comparisons were made for such resistance per equal weight of lint. The grader's decision as to fineness is really a decision as to hair strength, while his decision as to strength is largely a decision as to the uniformity of strength from fibre to fibre. Into this latter decision there enter other considerations, but it will suffice to leave the matter at this point, until one of the Graders shall also attempt to write a book on the subject. But it might be mentioned that the slipperiness of the lint has to be considered; it is almost impossible to break tufts of very "strong" fine cottons, because the hands cannot hold them firmly enough, but if the ends of the tuft are fastened with sealing-wax they can be broken with no more difficulty than any other cotton. Such slipperiness is partly due to the fineness of the individual fibres, but much more, in all probability, to the uniformity and frequency of the twist. Whatever the ultimate analysis may show, it is quite clear that the grader of cotton by hand is necessarily integrating a number of separate things under the name of strength, and we can now begin to see why graders' opinions are not always realized in the spinning-mill.

Interpretations of Grader's Terms.

In addition to this, it may be well to call attention to the fact that many of the features on which the grader bases his opinion are associative. Certain features of a cotton sample indicate that the crop was badly cultivated, and therefore the sample will possess certain other features. The substitution of pure-strain lint for commercial lint has

Familiar Associations of Features.

consequently some curious effects upon the grader. If the pure-strain lint has not been cultivated particularly well, it will show such "uglinesses" as would, in commercial variety samples, be necessarily associated with such irregularities of length and strength as a pure strain cannot show; the result is that the spinning-mill will give far better results than the grader would imagine possible.

Possible Mistakes in Grading.

Ignorance of this fact led to some serious consequences within the author's experience, and it would be well that those who may be concerned with the production of pure-strain cottons in the future should be aware of the risk.

Yarn Strength.—This is the only strength that really matters. Cotton is grown to be spun. If it spins well it is good cotton; if ill, bad.

We now meet with a new set of questions, many of which are still unanswerable in any exact expression; and although they have been dealt with more fully than other parts of our subject by previous authors, it may be well to present the case afresh. The main considera-

Unimportance of Hair Strength.

tion is this: that the strength of yarn has very little to do with the hair strength as determined by breaking strain, only about a quarter of the available tensile strength being realized. The strength of yarn is thus almost entirely dependent on the hold which the individual hairs take upon one another. Yarn does not break primarily through rupture of hairs, but through slip of hair on hair. Strength of yarn, within limits, follows the amount of twist which is put into it, even the variation in strength of yarn after

PLATE X.—AN IDEAL COTTON PLANT.

Photographs taken on July 25th (left), and August 27th (right), after several bolls had been picked. Giza, 1913, wide-sown. Christened the "Hedgehog Cotton" owing to the spiky appearance of the abundant floral leaves. This strain represents a rare combination of inherited factors, finally obtained in the fourth generation of a cross between two pure strains of Egyptian. There are no vegetative branches, all the stem joints are short, and the main stem ceases to grow at an early date. The result is a plant loaded with bolls, relatively free from shedding troubles. The lint of this strain is of the best Afifi type, and the author regards this as the ideal cotton-producing "machine." If subjected to natural crossing the stock would be entirely obliterated in a very few years.

mercerizing being accounted for in this way. If, therefore, yarn of a certain count and number of twists per inch is desired, the strength of it will depend almost entirely on the properties of the cotton employed, in respect to the grip which each hair takes on its neighbour. The author has practically no evidence to offer of the kind presented in respect of other characteristics, and our discussion must consequently be very general.

Uniformity.

In the first place, it is obvious that uniformity from fibre to fibre is a prime essential, on account of the weakest link. Uniformity in length must be of some importance, uniformity in diameter still more so, with uniformity in fineness, and probably uniformity in twist is the most important of all. Of these four features, the diameter is usually fairly constant; the length varies from the causes we have already discussed, and if its variation is excessive it can be regularized at the card or by combing; variation in fineness we have also discussed under the title of breaking strain, or hair strength, and there thus remains one important component—the twist of the lint hair.

Since the word "twist" has its special meanings in this connection, it may be advisable henceforward to refer to the twist of the fibre as "convolutions," in order to avoid confusion.

Convolutions of the Lint Hair

It was pointed out in Chapter III. on the Development of the Boll that the convolutions were caused by the presence of simple pits in the thickened wall, and were thus due to a definite structural cause, and not to any mystic gyrations of the protoplasm in the dying cell. This brings the whole question of these variations in the convolutions of lint hairs into line with other botanical investigations on pits in cell walls, and, although there are practically no data of the kind we require which are available at the moment, we shall be able to utilize them when they are obtained with other plants. The ideal cotton sample is one in which all the hairs are of the same length, diameter, and wall thickness, while all have the same number of convolutions per fibre in the same direction, spaced at equal intervals from end to end. Such a sample would interlock in spinning so as to give the maximum resistance to slip for whatever twist it received.

Reversal of Convolutions.

In the first place, the convolutions of the fibre do not always run in the same direction, but the direction reverses at intervals. It is not easy to say whether this is an advantage or not, under present conditions of spinning, for it should be remembered that, if all fibres were similarly convoluted, they would always have to be spun with the same direction of twist, if the maximum strength were desired. As to the causes of this reversal in twist, we can only conjecture that they result from some check in growth, taking place about the thirtieth day of boll development, when the secondary thickening is beginning, leading to irregular differentiation of the pit areas on the wall. They cannot be determined later than this, since otherwise the pits would not be formed through the whole thickness of the wall. They might possibly be determined earlier.

The spacing of the convolutions from end to end of the fibre would be determined at the same time as their direction, and in the same way.

In the third place we have to consider the "pitch" of the convolutions, which is of high importance. If the pitch is too low the fibres cannot interlock properly, and if too high they will be liable to "snarl" in preparation for spinning. At first sight it would appear that the pitch would be entirely dependent on the angle at which the pits were set to the axis of the cell, and this is undoubtedly an important component, as further investigations on the wild-cottons may show.

Uniformity and Pitch of Convolutions.

In addition, however, the thickness of the wall affects the convolutions, as can be easily realized on considering two extreme cases. If a lint hair has an extremely thick wall, so that the central cavity is practically obliterated, it cannot shrivel on drying, and therefore cannot form visible surface convolutions, in spite of the presence of the pits; if the same kind of hair, with pits set at the same angle, is very slightly thickened, the collapse on drying will be at its maximum, and the convolutions will be entirely determined by the pitting. Thus intermediate stages will have intermediate pitch in their convolutions. Regarding the question in this way we bring the convolutions into line with other properties of the lint which we have studied in their development. Convolutions are primarily determined by the angle at which the pits are set—probably an inherited character—and are modified from this by

Pitch and Hair Strength.

the extent and regularity of the thickening of the wall, excessive thickening reducing the pitch of the convolutions. The pitch of the convolutions which gives the best result from any variety of cotton in any particular class of spinning operations is in all probability definite for that class.

It would seem, therefore, justifiable to assume that although no precise data of the kind we require have ever been obtained as to the varying development of the convolutions, yet in all probability the same considerations apply as in the case of length and hair strength. All the factors, constitutional or environmental, which modify the latter from ideal uniformity, also modify the convolutions in a parallel way.

Uniformity.—Throughout the whole of our discussion of the development of raw cotton there has been one recurrent ideal, namely, the production of uniform cotton. By various stages of analysis we have seen how it is possible to attain an approximation to this ideal, and how the difficulties may all be overcome—if it pays to do so—with the exception of such irregularities as are due to the "struggle for existence" between individual hairs on the same seed.

Perfect Cotton Unattainable.

The fact that we have had to carry our analysis into such minute details, and to link these up with such remote causes, shows very clearly that, however interesting the knowledge may be, a strictly uniform sample of cotton-lint can never be grown.

Equally, however, our analysis shows how very far

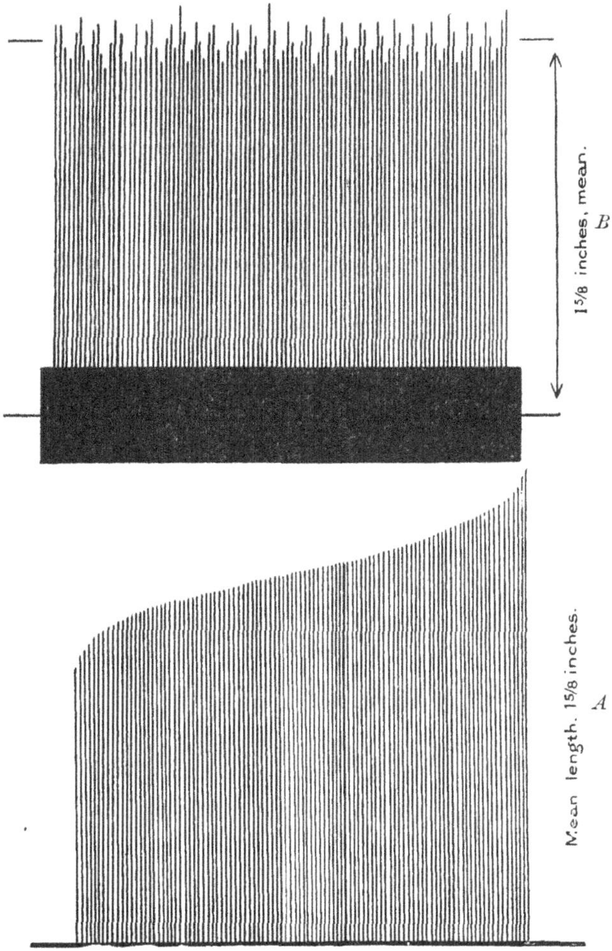

FIG. 18.—LINT LENGTH OF A GOOD SAMPLE.

A represents the lengths of 100 lint hairs taken at random, and measured in a good commercial sample of Sea Island by O'Neill in 1863 (computed from O'Neill's figures), and checked by the author on samples of Egyptian varieties graded as "excellent" for regularity.

B, the same hairs as *A*, but mixed together and held in a clamp, to represent the grader's inspection in hand-pulling a sample. A sample which actually shows this amount of irregularity when sorted hair by hair is considered excellent, and pulls with a "hard edge."

remote from the practicable possibilities the best samples of to-day must be (Fig. 18). It is hard to believe this is so when one is handling some specially good St. Vincent cotton, or similar raw material; but nevertheless it is quite certain that the best cotton grown to-day is far from reaching the moderate uniformity which it would be quite practicable to attain by comparatively slight refinements in seed-supply and cultivation.

In the light of our analysis it is almost impossible to say what uniformity does *not* mean. Practically every property which a commercial sample possesses is partly the property itself, and partly uniformity of that property. Graders' strength is partly uniformity in strength, lustre is partly uniformity in cell-wall formation; deteriorated colour is non-uniform colour, and so forth. It is to be hoped that, even if the class of investigations described in this book has no direct applicability, they will at least facilitate the interpretation of these terms of common speech into their real components.

Ubiquity of Uniformity.

CHAPTER VI

THE DEVELOPMENT OF COTTON-GROWING

THERE is some peculiar fascination about cotton which defies analysis. Probably the enormous size of the industry, and the unsuspected revelations which it continually makes to the student, have something to do with its charm; the gap between the native cultivator and the mill-hand is so wide that any person dealing with any part of the cotton trade must of necessity take some interest in the other parts which only concern him remotely. It is a humiliating reflection on the intellectual effect of prosperity that the relations between science and the cotton trade became steadily less intimate during the past century, so that the sentiments expressed by writers on the subject read as if they had been written backwards, the writers of a hundred years ago summoning all the scientific knowledge at their disposal, while those of to-day frankly admit that the trade has been working by rule of thumb

The obvious retort is that science was found a fallacious guide, and that the only path worth pursuing was the one which led to financial profit quickly. That the latter was the case till the end of last century is undeniable; the world had all the cotton it wanted for the time being, and no recondite study was necessary. Since A.D. 1900 the situation has altered enormously. Fears of dependence on the American crop in the event of a shortage, the developing uses for strong cotton fabrics in modern light and rapid instruments of locomotion, and the increasing accessibility of the colonies, have all led to initial steps in the development and control of raw cotton supplies, and to their better utilization when obtained.

Science.

To outline some of the principal ways in which these developments can most efficiently and rapidly be made should be of some use, if it be remembered that the suggestions are based merely on the author's personal opinion. The greater part of such development work is as yet but a stumbling attack on the problems involved, wasteful of time and money. If natural science is to take any concern in economic affairs, it should at least be able to offer some suggestions of a general nature which might facilitate the work of those who are clearing jungle, digging canals in the desert, and coping with the difficulties of administration, in order that more and better cotton may be swallowed up by the bale-breakers in the mills.

Finance.

In the first place, it should be clearly understood that science must follow the financier in the first instance. If transport, labour, water, are only to be obtained at a high price, cotton-growing must be a failure as a business proposition. Moreover, the scientific economist must follow close upon the financier, for the hasty development of cotton-growing in new country may lead to many troubles; the price of native labour may be raised to a prohibitive figure, from which it will not recover for years. Sometimes the scientist may be able to take the matter in hand, and by improving the value of the raw material, or by reducing the cost of production, compensate for these disadvantages; but this statement has been made so often, and so seldom realized, that it might be better not to repeat it.

Essentially, then, cotton is a cheap-labour crop, and a hand-labour crop as well (Figs. 5 and 6), and will remain so until the dream of a mechanical means of picking has been realized. In this respect the author is inclined to think that the aim of inventors has ranged too far; a long experience of the routine repetition of operations, such as those on which the diagrams in this volume are based, has led him to value small refinements in method. The position in which a pencil is laid down by the side of the balance-case may make a difference of 10 per cent. in the number of weighings effected in an hour. Similarly, if a strip of bent tin or a curly piece of wire would enable the pickers to gather a few more bolls in the same time and with the same effort, it might make the difference between the success and failure of a new cotton-growing area. Even such trifles as dropping the load of picked bolls at the end of the row, instead of carrying it along other rows, make a difference in the amount picked, and some simple experiments of this kind might be quite usefully conducted, in order to ascertain how time might be economized, without extra labour from the operatives.

Mechanical Pickers.

In developing a new district for cotton-growing, provided that the economic situation is satisfactory, two things are necessary. Firstly, a crop has to be grown in the field, and not merely in gardens; secondly, the reasons for its failure have to be ascertained. The author's use of the word "failure" is deliberately designed to draw attention to an aspect of agriculture which has not been fully viewed before; every crop is more or less a failure. More usually we say that it is more or less of a success, but in the case of cotton it is almost better to express it the other way.

The Failure of Cotton.

So long as good land in Egypt can produce a normal crop of 700 pounds of lint to the acre, with an average of over 450 pounds for the country, even now, while the U.S.A. averages about 200 pounds, and India less than 100, we have a definite basis of comparison for the degree of failure which we call success in cotton-growing.

To return to our second essential, namely, a knowledge of the reasons for the comparative failure of the crop. It is obvious that a full knowledge of this kind can never be obtained, and that it will take years of research to provide even reasonably intimate knowledge; but the immediate demand may be put in a simpler form, to wit, "How did the crop behave?" On the surface this would seem to be a childishly simple question, but it is one which no cotton-grower could answer in any form giving the scientist information from which to draw conclusions. With some crops it is a relatively simple matter to describe how the yield was attained, but when the yield is being

The Behaviour of a Crop.

built up day by day over a period of two or three months, ordinary casual (or even skilled) observation breaks down.

The first step in developing cotton-growing for a new country should be the procuring of records showing how the yield was built up under optimum conditions. This involves also determination of the optimum conditions.

Experimental Trials.

The optimum conditions of the site and year in which the trials were made are very simply defined as those which produced the largest crop, which also —other things being equal—will be the best crop. The conditions which produce the best crop can only be ascertained by trial and error, but there are right and wrong ways of so doing; much of this work as done to-day consists of very little trial and very much error. Certain conditions of the environment can be controlled in the trials, while others cannot. Weather is uncontrollable as regards temperature, and only as regards water when there is no rain; in irrigated land the water-supply should be made one of the subjects of experiment. Although the weather cannot be changed, its incidence on the plant may be changed, by making the time of sowing another subject of experiment. The soil may be modified by manurial treatment, though experiment in such direction should be restricted initially to those methods which are likely to be practicable for the general crop. Lastly there is the arrangement of the plants on the area cultivated, which is of more importance than has been realized in the past; there can, for example, be little doubt that the American crop would be about 30 per cent. larger if it were not necessary to put the rows far apart so that horse-hoes could work between them.

Taking three controllable variables only—manure, sowing-time, and spacing—a good deal of most practical

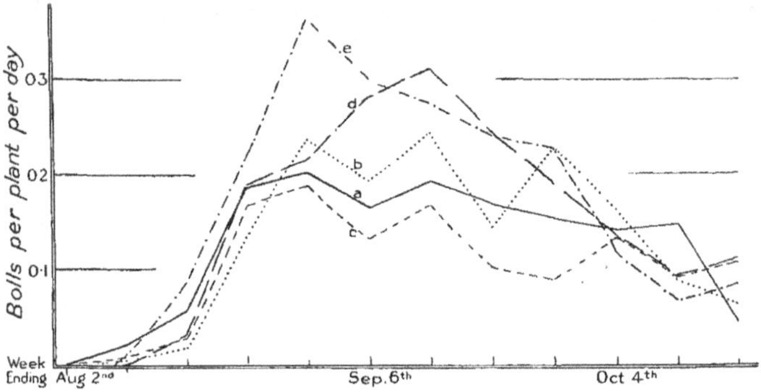

FIG. 19.—IDENTICAL PLOTS OF COTTON.

These curves show the number of bolls opening in each week of the season on five plots which were nominally exactly alike, in various parts of a single acre of land (shown in Pl. XI., XII.). Perfectly definite differences exist between them throughout the season, due to variation in the deeper layers of soil.
Fig. 2, p. 24, is constructed from the averages of such sets of five plots.
Variety, Domains Afifi; site, Giza, 1913.

MEAN YIELD = 400 POUNDS OF LINT PER ACRE.

Per plot:					
Total yield in pounds	372	379	303	450	496
Deviation { Actual	−28	−21	−97	+50	+96
Deviation { Per cent.	−7	−5	−24	+12	+24
Legend	———	………	- - - -	—— —	—·—·—

information can be obtained in a single year. It remains to consider how the optimum for each of these variables should be ascertained, and this brings us to the question of experimental plots.

PLATE XI.—CHESS-BOARD PLOTS

Showing the arrangement of fifty small plots at Giza on April 12th, 1913. The size of the plots can be seen from the three lying next to the road, which are conspicuous through having recently been watered. Each plot consists of ten ridges, each of fifteen pairs of plants, and bears a serial number at the top of a tall stake. The results from these particular plots are used in Figs. 2 and 19, pp. 24 and 158.

In the last five years we have seen a revolution in our knowledge of field experiments, the methods of conducting them, and the errors inherent in them. The two chief features of this new knowledge are as follows: The increased precision obtained by increased size of plot is practically negligible beyond about one-tenth of an acre; the errors in comparison of plots which are supposed to be identical is such that with English wheat the plots may differ as 84:116 by pure accident, and only half of them will be as closely similar as 95 : 105.

Small Plot Trials.

A great part of this error is due to soil variations which cannot be eliminated. When we deal with a deep-rooted plant like cotton, which sends its roots through more than two metres depth of soil in a season, the errors are much greater from this cause. Identical plots of cotton may differ by nearly 75:125 through normal accidents alone. and half the plots are bound to differ more than 93 : 107, this degree of difference being shown on total yield, and being proportionately more on separate pickings (Fig. 19).

Errors of Plot-Comparison.

Recognition of the existence of this very high degree of uncertainty in comparison will account for the uncertainty which attends on our present knowledge of the cotton crop. Practically the whole of the work of the past fifty years on experimental crops of cotton will have to be repeated in this new light, just as Mr. Leake has pointed out that nearly all the failures in introduction of new cottons into India have become devoid of significance in the light of

Need for Accurate Results.

recent discoveries about natural crossing. There is only one way in which certainty can be attained, and that way is a laborious one, though not at all impracticable ; by dividing the experimental area into small plots (Pl. XI.), putting five under each kind of treatment, and scattering these five over different parts of the area, the precision of the results with cotton may be increased to 9 : 11 as the maximum possible dissimilarity due to accident, when the five-plot averages are compared. It is frequently objected that the trouble and labour of handling small plots makes them impracticable; there is, however, no escape from the fact that only in this way can a reasonably correct answer be obtained; whether it is "practical" to obtain an answer which has practically no meaning, and "unpractical" to expend a little more on labour to obtain one which has a definable significance, must be left to the future to decide. It should be noted that the same amount of land and ordinary cultivation is required in both cases. The additional trouble is in the handling of the small plots, and especially in laying them out and sowing them. One working suggestion in the former respect may be useful, namely, that no attempt should be made to differentiate the plots in the field-work; each plot should have a serial number, be observed and treated under that number, and the final grouping effected only when working up the results of the observations. Working in this way it is not impracticable to combine two, or even three, experiments into one; thus sowing-time and spacing could be handled in one series of plots in the following way:

The Handling of Small Plots.

Four acres of land cut up into 125 plots, twenty-five being sown each week for five weeks grouped around the probable date of sowing, with five different spacings of five plots each. The data thus obtained could be taken as being correct within the 9 : 11 extremes for spacing and sowing combined, and for either spacing or sowing separately their extreme possibility of error would be 95 : 105. Whether such combined experiments were practicable, or whether each point would be investigated separately, would depend solely on the labour available. This class of experimental work requires either careful supervision at the times of sowing and picking, or else the training of a few natives to act as observers, with a modicum of thought in the arrangement of fool-proof methods for them to follow in making the observations.

By conducting experiments on these lines, so as to obtain a reliable answer, much time and money is economized in the following year, when the results from the plots are applied on a larger scale. The case of sowing-time in Egypt is very much to the point; the author has shown that the early sowing of cotton before a certain date is of no advantage, and may bring a loss, while sowing after that date delays maturity; the cause of the existence of this "critical date" would appear to lie in the temperature of the soil, which at depths of a foot or two undergoes practically the same seasonal changes in temperature every year.

Utility of Accurate Results.

So far we have sketched the methods only by which

the optimum conditions of cultivation for a certain site and year can be determined. It still remains to answer the question as to how the plants behaved under those conditions?

The cotton crop is thoroughly misleading in its appearance. A field may appear to have but little ripe cotton in it, and yet be full of open bolls which are obscured by the leaves. Another field may appear to be flowering profusely, and yet be about to stop flowering almost entirely. The best experts of those who spend their time in travelling about a cotton district and reporting on the crop condition are well aware, and will admit, that they cannot estimate the yield of cotton ripe in any particular field to within 10 per cent. Yet there is scarcely any crop which has been so entirely discussed on purely general ideas as to its appearance.

Uncertainty of Subjective Opinions.

The question arises as to whether any more accurate data are worth obtaining in the early stages of development, and the answer is most certainly affirmative. Take, for example, the case of a rain-storm occurring at an unusual time in the first season of experimental work; it is desirable to know whether the results obtained in that year are still generally applicable to future years, or whether they have been entirely falsified by the storm. To justify expenditure on elaborate small-plot experiments, it must be shown that their results can be made of general significance, and applied forthwith to other possible sites and seasons.

Elimination of Accidents.

Fortunately it is a tolerably simple matter to obtain such records, given a few native labourers and a little training. The observations may be taken on the entire plots, or on groups of 200 plants each, counted and marked off by stakes in each plot. The observations which are practicable are those of bolling and flowering, and the object of these observations, as well as of others which are less easily obtained, is to present a continuous record of the behaviour of the plants. The ideal would be to take these records daily (Figs. 4 and 10), but this would rarely be practicable in early development work, and slightly longer intervals may be substituted.

Crop Records.

The records of bolling are obtained by picking or counting the number of bolls open on the observed groups of plants each week, so as to obtain the number ripening in each week per plant or per plot (Figs. 10, 13, and 16). When a large number of plots is being handled, it may be convenient to take one day of the week for one series, and one for another, but adhering strictly to the same day for each series. The figures may be expressed as the number of bolls ripening, or the weight thereof, or—best of all—both ways; if both are taken, the average boll weight each week is thus obtained, which is an important consideration. By plotting the results on squared paper, it is easy to see, not only which plots gave the best yield, but which were earliest, etc., and to deduce from these curves the reasons for failures.

Bolling Records.

The flowering records (Figs. 2 and 10) are even more

valuable, but they necessitate daily observations, or alternate days at the least, if daily observations are absolutely impossible. They are more valuable because they are not subject to so many sources of error as the bolling records, and therefore give more accurate comparisons from plot to plot, especially as regards the early stages of growth. After the flower has opened, it may be prevented from ripening into a boll through shedding caused by water shortage or excess of water, or by weather; or through the attacks of insect pests or fungi. The bolling record thus merely shows how the crop was produced, but the flowering record helps to explain the why and wherefore. Flowering records cannot be taken satisfactorily on odd days, or even at regular intervals, because the flowers do not accumulate as the bolls do, and also because the rate of flowering varies very greatly from day to day, owing to previous variations in the growth-rate of the flowering branches. In spite of these disadvantages, if flowering records can possibly be obtained, they are well worth the trouble, on account of the insight they give into the way in which the yield was formed. It is not always realized that the number of flowers opening is the main determinant of the final yield of a cotton-field; the yield may be less than the flowering would indicate (Fig. 10), but it cannot be more. The results can be plotted similarly into curves, on the same scale as the bolling records, per plant or per area; the difference between the two curves shows the loss on any day or in any week from shedding and from insects, and

Flowering Records.

Flowers and the Yield.

the departure of the flowering curve from a theoretical form gives evidence as to the nature and magnitude of the causes affecting the plants.

The utility of these Plant-Development Curves is almost endless, as the author has shown in Egypt. To study the data obtained in ordinary field experiments, after being accustomed to using these continuous records, is like making use of a dictionary from which many pages have been torn away. The difference between them and the ordinary data for three pickings is similar to the difference between the inked trace of a barograph and the guesses founded on the tapping of the barometer. The trouble of obtaining them is appreciable, but the cost is trifling; one experiment with 100 plots on 2 acres, conducted by the author, in which many more data were taken than those sketched above, cost £30 more than the ordinary cost, for all observation salaries, stakes, headman's time, and clerical appliances. Three sets of plots of fifty each, directed to the examination of ten different spacings, sowing-times, and manurial arrangements, should, a year later, repay the outlay upon them many times over.

Use of Crop Records.

A modification of these methods may be used, with certain limitations, for testing varieties or strains when only very small amounts of seed are available. A field of ordinary crop is taken, and in it are sown rows of the seed to be tested, replacing the ordinary crop seed which should have occupied the same places. These rows consist of about a hundred holes each, and not less than five such rows,

Variety Testing.

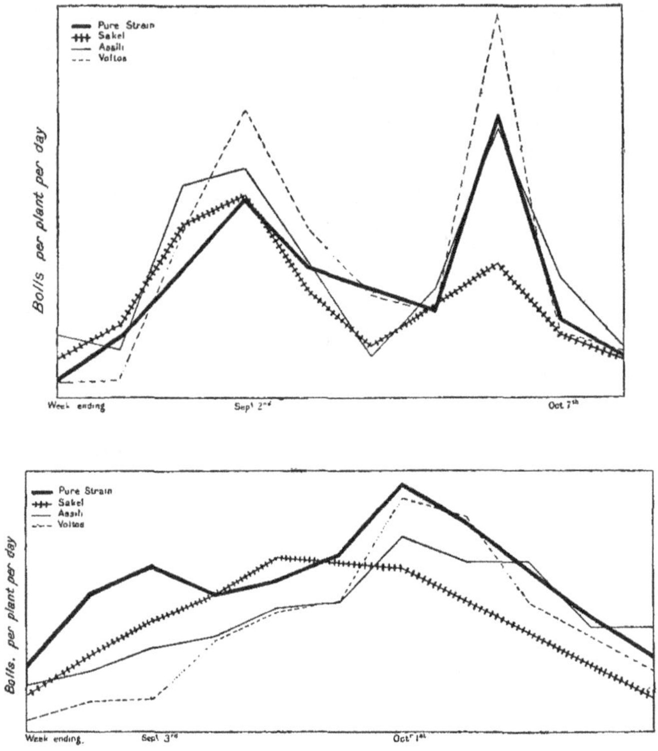

FIG. 20.—VARIETY TESTING BY BOLLING CURVES.

Illustrating the use of Plant-Development Curves for the comparison of varieties, using very small amounts of seed sown in scattered groups amongst ordinary field crop.

The 1911 comparison was made on land which had borne three successive crops of cotton, but was well cultivated.

The 1912 comparison was one richer land, and the curves consequently rise higher, but irrigation was delayed in July, causing shedding of the flowers, which consequently reduced the yield for a while in September.

In both years the slight lateness of the Voltos variety stands out, while Sakel clearly cannot tolerate water-shortage.

JUNE 25.

JULY 21.

PLATE XII.—THE DEVELOPMENT OF A FIELD OF COTTON.

A portion of the same view as in Plate XI., showing how the soil becomes shaded by the densely crowded foliage.

preferably more, are scattered about the area. If more than one kind is to be tested, parallel and adjacent rows are used. Records are taken from these rows, showing their flowering and bolling, and from similar parallel rows of the ordinary field crop amongst which they are sown.

In this way it is possible to obtain remarkably exact comparisons when only a few ounces of seed are available, and a very marked economy may be effected in the following year when full field trials are undertaken, since not only have the useless varieties been eliminated, but valuable information about varietal peculiarities has been obtained (Fig. 20). Thus the accidental failure of a new variety can be distinguished from a real failure.

Meanwhile there is the question as to whether it is worth while attempting seed-breeding, and this is problematical. Seed-supply is only worth doing when it can be done very well, and a new country is rarely suitable for refinements of this kind, although it may be quite practicable to get minute data of the kind we have discussed from the land near a residence.

At the same time it would be well worth while attempting to make pure strains from the commercial varieties which were most successful, ultimately replacing these latter by them.

Pure-Strain Production.

The isolation of pure strains of cotton is another of the many commonplaces which the public likes to enshroud in mystery. There is nothing mysterious in the process; it is almost true to say that no skill is required, nor any knowledge of botany, nor even of cotton. The sole essential is ceaseless, unflagging, searching accuracy in handling the material (Pl. I., VI., X., XIII., XIV.).

Roughly summarized, but with most rigid definition of every word, it consists in obtaining seed from single plants by self-fertilization exclusively, until plants are found which give offspring all exactly alike constitutionally in every visible and measurable feature.

There seems to be some conviction at the back of many minds that a new kind of cotton can only be obtained by multiple crossing, destruction of organic stability, increase of tendency to reversion, interference with the balance of nature, and consequent aftermaths of like vagueness. Actually the production of pure strains is as straightforward and definite a process as the separation of sugar from sand.

The cost of such work is high, however, even if the actual purification research is not charged to it. To maintain a single pure strain from year to year, avoiding all contamination by crossing and mixture, cannot be done at a cost of less than about £50 per annum for a renewal supply of only 20 kilogrammes of seed, merely for appliances (Pl. XIII., XIV.), and without counting the cost of skilled labour. It would be to the interest of new countries to develop pure-strain cultivation as quickly as possible, when the old countries have strains to spare, for many such strains will be isolated and tested and found to be slightly unsuitable for the country of origin, and yet might be perfectly suitable for some other country. One of the coming features of the cotton trade in this respect will be an

Development of Pure-Strain Breeding

PLATE XIII.—MAINTENANCE AND PROPAGATION OF PURE STRAINS.

Bee-proof cages of brass gauze mounted in sectional panels of iron, each cage being one-thirtieth of an acre in area, costing about £70, and producing about twenty pounds weight of uncontaminated seed each year for the purpose of sowing propagation plots. This method is actually cheaper than bagging flowers by hand for the same purpose. View of, and through, the corner of two cages.

PLATE XIV.—A BIRD'S-EYE VIEW OF SMALL EXPERIMENTAL BEE-PROOF CAGES.

International Seed Register and Bureau, so that pure strains when once isolated shall not be left to die, but shall be kept alive in small quantities of seed, with accurate published descriptions of their performances in the conditions under which they have been tested, and shall be available for multiplication and further testing in any other country. The cost of such an organization will be borne by the trade as a whole, or by the consuming side of it alone, since it will be to the advantage of the spinner, and not to that of the successful grower, that such an organization should exist. At present not more than one per cent. of the work done on plant-breeding remains economically available. At the same time such an organization would have to be run very strictly, nothing but statistical evidence being admitted, either for purity, cropping capacity, or spinning properties.

This last brings us to another probable development of the future. At present it is very difficult to ascertain what is the comparative value of any sample of cotton, since there is no means of testing raw cotton. We have seen in the preceding chapter that the only test of value is the test of spinning, and some persons have suggested that miniature spinning-machine testers might be practicable. This is highly improbable, if not actually impossible, and in default of any existing indirect methods of analysis the cotton must be put through ordinary standard machinery. It is but rarely that any investigator has the good-fortune to have the courtesy extended to him in this respect which the author has received from the Fine Spinners' Association.

Spinner's Testing-House.

The trouble of conducting these tests in an ordinary mill is very considerable, but it should be practicable to establish a "Spinning Testing-House" for raw cotton, as suggested by Mr. J. W. McConnel, in which sets of machinery were installed for the special purpose of handling ten-pound samples. The fee would necessarily be fairly high, but the use of such an institution would not be confined solely to growers. The tests would have to be standardized for a range of counts and classes of yarn, and the results of the tests presented statistically as far as possible. An immense amount of uncertainty would thus be eliminated from the grower's work, and a series of standard records would accumulate.

It is interesting to look back from the present day to the results obtained by the first of the author's predecessors, Mr. O'Neill. His papers were read in Lancashire in 1863, and from them we can make certain comparisons with the cottons of the present day. The intervening fifty years have heard much talk of progress, and have seen many extensions of cotton-growing areas. Nevertheless, the good cottons which Mr. O'Neill handled were every whit as regular and good as those of the present day. Modern civilization has scarcely begun to affect the cotton-plant.

Fifty Years Ago.

This book has been written in the hope of clarifying ideas on the subject, and facilitating further inquiry, so that the cottons of fifty years hence may be more dependable than those of to-day.

PLATE XV.—{ SACKS OF SEED-COTTON, LOWER EGYPT.
BALES OF LINT, LOWER EGYPT.

APPENDIX I

METHODS OF INVESTIGATION

THIS appendix is intended rather for the use of those who wish to re-investigate this work. The subject is the methods of investigation employed in recording and controlling the condition of the plants, in the treatment of the developing bolls, in the examination of the seed-cotton, and in the presentation and analysis of the results.

Crop-Records.—A system of routine records of the daily condition of the crop was developed by the author, in the first instance for individual plants in breeding experiments, and later for field crop conditions. At the end of the author's service in Egypt these records had become so comprehensive that they gave information as to the state of the crop of the whole country, even when taken on a single site only. They consisted in measurements of the daily growth, of flowering, and of bolling, with subsidiary measurements of other features, such as shedding, and, in fact, of any character of the plant which the exigencies of research and convenience combined to render worth measurement. The cost of obtaining these data was slight, unskilled native labour being trained for the purpose, and checks imposed ultimately by the methods themselves. The headman checked the work of his subordinates, and very little supervision was required by the author or his assistant, since every set of plants recorded constituted a check on the exactness of the records from other sets. If the curves presenting the data from two identical sets of plots did not coincide after computation each day, slackness on the part of the Observers was suspected at once, and it is only fair to say that such occurrences were very rare. The Observers could not possibly concoct their results, since they had only a vague notion of the nature of the dozens of different groups of plants which they were observing, and the actual numerical records meant nothing even to us, until they had been grouped, totalled, divided, and plotted on squared paper. Our records of crop condition at Giza in 1913 attained nearly to absolute precision day by day from May to November.

All data were taken from definite groups of plants called "observation rows," containing a known number of individuals, and the results were all computed down to terms of an average plant. Presentation of the condition of such a plant as should be the average for a whole field, or even larger areas, was obtained by sampling various portions with a regular "scatter" of such "observation rows." Any one of these rows could be chosen for the source of such material as is required for investigating the development of the lint, whether for pickling or for testing, and reference to the routine records would show at a glance the daily rate of growth in tenths of millimetres, the fractional number of average daily flowers or bolls, dates of irrigation, and so forth; while against these could be plotted the corresponding data of each day for temperatures, sunshine, humidity, wind, evaporation, soil-water content, subsoil-water level, or any other data which might be relevant to the subject under investigation.

Presenting the state of the crop thus in the form of an ideal average plant at once brings the agricultural problems into directly botanical form, or, rather, propounds those problems in terms of plant physiology. With such exact records it is possible to trace most minute differences, of which one example may suffice:

In former years the author had ascertained the period of boll maturation for several varieties and strains by marking open flowers, and watching for their date of opening. Among these was a pure strain, No. 77, which in 1910 had a maturation period of forty-eight days, with a Probable Error of 3 per cent. In 1913 material was taken for these lint-development investigations from this strain, but no direct record of the maturation period was repeated; the question arose as to whether the change of plot and of year had altered the maturation period, or whether forty-eight days could still be taken as correct. A full set of daily observations of flowering and bolling upon the variety "Domains Afifi" were smoothed to five-day means, and the two curves superposed; they fitted with an interval of fifty-one days. The process was repeated with the relatively imperfect data from the observed group of No. 77, and indicated a slightly shorter period; then the dates of irrigation of the group of No. 77 were taken, fifty-one days added to them, and the dates thus obtained were marked on the bolling curve, when it was found that each date was three days later than a sudden rise in the bolling, which we knew to be due to diminished shedding; this marked the interval between the opening of the flower and the opening of the boll as 51 – 3 (or 48) days for strain No. 77 in 1913, repeating the conclusion formed by direct determination in 1910. A system of records obtained at very little personal trouble or official cost, which will thus cast up evidence at will, splitting apart differences of three days in a period of seven weeks, is obviously not without its uses.

Such an example also shows the precision to which field crop botany might be developed.

A fuller account has been given elsewhere of the methods by which these records were obtained, and only sufficient has been said here to explain the nature and purpose of them, and to make the appearance of the plant-graphs in the figures intelligible.

Cytological Methods.—The earlier part of this work was effected in 1905 and 1906, when technical difficulties and the absence of physiological data made it plain that no further real advance was possible until more had been found out concerning the physiology of Egyptian cotton. The indirect attack, or flanking movement, took years to develop, but in 1912 enough was known to justify a resumption of the direct cytological investigation, and material was collected which cleared up the whole story provisionally in about a week at the microscope, and was completely confirmed by the parallel and subsequent studies of cotton ripened from dated flowers and bolls.

For the later work we labelled batches of some 200 flowers on pure-strain families on a single day, and collected from these some three or four bolls every third day afterwards until maturity was complete. These were incised in each carpel wall, and pickled directly in acetic absolute. Most of the examination of this material was done in glycerine jelly, and without staining, though fuller methods were also employed. The technical difficulty of combing out long fibres devoid of any secondary thickening, so as to extricate them from the tangle of lint without breaking them, was solved by the simple plan of combing with a small-tooth comb in the usual way, but in warm water instead of in air, after pickling.

The microscope employed was the large Zeiss stand, with condenser, compensating oculars 2, 4, 12, and 18, objectives *a*2, A, D, and $\frac{1}{12}$ oil immersion, and the 3 millimetre apochromatic objective, all by Zeiss.

Sections were cut by hand on the hand microtome or on the Cambridge Rocker, embedding in 60° C. paraffin. The more strictly cytological material of the early investigations was fixed chiefly in strong Fleming, and stained chiefly with Heidenhain's hæmotoxylin. Some of the 1912 material was also examined with these reagents.

Sectioning of the lint itself deserves some additional mention, as the difficulties which the author encountered have caused him to wonder how the sections so freely figured in other works were obtained, and a special method had to be worked out to cope with them in the 1912 material. Part of these difficulties was due to the stroke of the Cambridge rocking microtome, which does not give a drawing cut, and is consequently helpless against cotton fibres, even when the paraffin is cooled to 0° C., these fibres being equal in tensile strength to wrought iron, much more elastic, and only 0·018 mm. in their widest sectional diameter. Tolerably satisfactory sections were obtained with a drawing cut on the Swift's hand microtome, once satisfactory embedding had been attained.

Embedding was found to be most difficult. Ordinary xylol infiltration, even in three-day steps, was useless, and finally success was attained with chloroform *in vacuo* at 100° C., with the added advantage of very rapid handling. Material which had been left in ordinary alcohol overnight could be in section under the microscope by noon, all operations being conducted in test-tubes connected with the vacuum water-pump by tubes passing through the thermometer hole in the roof of the water-jacketed drying oven, saving the expense of a vacuum embedding bath. The stages were: alcohol, absolute, absolute - chloroform equal, chloroform three times, chloroform-paraffin 60° C., paraffin 60° C. twice. In this way the lumen of the cell was thoroughly infiltrated, and clean sections could be cut, cemented with albumen-glycerine, and handled as smears.

Preparation of Dated Samples.—Samples of known history, upon which the physiological hypotheses based on the cytological evidence could be tested, were most simply and completely obtained by taking daily pickings. The result is material which has undergone an endless variety of environmental experiences, these latter being recorded in the routine records already described. Some severe maltreatment of the plants may be advisable in the first experiments, in order to provide an unmistakable effect as a base-line. The result is that such material provides not one experiment, but a series of some sixty experiments in one, with consequent economy of time and labour in handling the material.

Continuity of Record.—Such treatment involves a principle of considerable moment to biological research, and especially in agricultural matters — namely, the utility of continuity in data. Every additional point in a curve increases the definition of the curve, and a continuous projection is, moreover, much closer akin to the requirements of the practical man than the exact definition of a few isolated points. Much of the agricultural investigation of the past has been unconsciously an effort to draw complicated curves from the knowledge of only one or two points along their course, so that divergencies of opinion have naturally arisen—and have given a proverbially bad name to agricultural experts.

Continuous data are so much more easily interpreted, especially in field records. We are all familiar with the recording barograph, and find it much easier to observe the changes of the weather from the rise and fall of its inked trace than in the days when the old game of tapping the barometer prevailed in our households; the barometer might have fallen ½ inch and returned again during the night, but we were none the wiser. The physical laboratories think in continuous projection, but in many fields of biological work even the effort to obtain continuous data has not yet begun, especially in tropical agriculture.

Until the author had familiarized Egyptian workers in agriculture with his "bolling curves," even the best-informed of them were under the impression that the three "pickings" of cotton-fields corresponded to three separate orgasms of energy on the part of the plant; in point of fact there is some accidental justification for the origin of this belief, but the belief long outlived the circumstances which gave rise to it.

This idea of converting our fragmentary knowledge into a continous sequence was in the author's mind at the very beginning of his Egyptian work, but not definitely formulated in working methods, excepting for a conviction that, because much work on the cotton crop in bulk had failed to yield many generalizations, it would be worth while to work in the opposite way, by studying a few plants carefully, and from this—as methods developed to make possible the study of many plants carefully—the drift towards continuity became apparent.

The practical objection to developing continuous records is that, unless assistance in some form is available, there can be no day of rest for the observer, for a single day or period omitted from the records spoils two intervals, the one before and the one after.

The collection of bolls on successive days is not entirely satisfactory, since there is a slight subjectivity involved in deciding whether a boll is fully open or not; and unless the same observer can always collect the material each day, this may lead to slight irregularities, though only of a single day in either direction, the fully open bolls being fairly definite. More precision can be obtained by labelling flowers on successive days, and collecting the bolls which ripen from them some seven weeks later. In this case the boll can be left on the plant as long as may be convenient, but the method involves more trouble and risk of mistakes in identification.

In either case absolute exactitude is unattainable, since, although the maturation period is definite (*vide supra*) for any given strain, it also has a definite range of fluctuation from accidental circumstances acting on the individual boll or plant. The probable error of the maturation period is 3 per cent. in Egypt; this on forty-eight days is one and a half days, or, in other words, half the bolls observed will mature between forty-six and a half and forty-nine and a half days, and no ordinary accident can make the boll take more than fifty-three or less than forty-three days. Therefore, even if we define either end of the maturation of the particular boll with complete precision, we cannot be quite sure of the age of the boll at the other end on any given day; we can, however, define exactly what the chances are, and from this we can deduce the size of sample necessary to smooth out these accidents. In practice a ten-boll sample is quite satisfactory, and fifty bolls are ample.

The seed-cotton so collected is weighed to determine the average weight of the boll-content, combed to determine lint length, and ginned with concurrent determination of the ginning out-turn and the mean seed weight. Subsequently the lint is graded, and tested for breaking strain. The numerical statements thus obtained cover a great part of the results described in the text of this book. Other determinations, such as diameter of the lint hair, weight of a centimetre of the lint hair, thickness of the wall, amount of twist, etc., have been made only on samples chosen as typical of particular sets of circumstances.

Ginning.—The samples were ginned on a 12-inch roller gin made by Platt Brothers of Oldham. The author is indebted to this well-known firm for the very courteous loan of one of their 4-inch "Missionary" gins to use in England, while awaiting the arrival of a little Churka gin from India, but these two latter implements were only used for a small part of the work here described.

The 12-inch gin was run at about twenty-five revolutions of the roller per minute by hand, this slow speed being employed in order not to damage the fibre unnecessarily; in the ordinary way these hand-gins depreciate the quality of fine cotton very noticeably. For small samples the little Churka gin, or its modification with a twist-geared iron upper roller, is extremely useful, being so easily cleaned, and for work of this class it would be worth while for any investigator to make, or have made, a rather larger and improved form of this implement. The ordinary roller gin necessitates some troublesome attention to insure that the ginning out-turn is correctly obtained, since lint coming off the roller is liable to be lost by coiling round the roller axle. In any case a canvas shoot should be mounted on a light frame which is clipped under the beater-blade frame, in order to catch the seed, which otherwise has to be picked out of the interstices of the machine.

Ginning Out-turn.—The determination of this ratio of lint to seed-cotton, which we now know to be specific for each strain, just as lint length is, with fluctuations, is best effected by weighing the seed-cotton, and then the lint ginned from it.

From the figure thus obtained, together with the mean seed weight, one can obtain the weight of lint per seed by computation. Determinations by weighings of the seed are unsatisfactory, as a small percentage of seeds are always lost in the act of ginning, though the use of a canvas shoot and a loose overhead cover reduces this loss to about 1 per cent. with the 12-inch gin running as fast as two men can turn it.

The weighings and computation (by slide-rule) required to obtain the ginning out-turn can be done in about five minutes per sample. If the seed-cotton is weighed out against a fixed weight of 20, 50, or 100 grammes, there is then only one weighing to be done with loose weights, and the out-turn can be calculated by mental arithmetic without much risk of making a mistake. This reduces the time required considerably, but it still requires three or four minutes per sample. By a simple modification of the steelyard, provisionally patented as the "Slide-Rule Balance," this time was cut down to less than one minute, which is a matter of no little importance when hundreds of samples have to be dealt with. This balance works with two riders and two pans, movement of the primary rider bringing it into equilibrium with the seed-cotton on one pan, and movement of the secondary rider on the unaltered primary bringing the lint into subsequent equilibrium on the other pan. The primary rider is so graduated that the position of the secondary rider upon it shows the percentage of the second weighing to the first weighing, and thus there is only one figure to write down—namely, the ginning out-turn as desired. A simpler form with one rider operates for a fixed weight of seed-cotton only.

Seed Weight.—Speed and accuracy in determining the average weight of the seed in a sample are best obtained by weighing out 10 grammes of seed, and subsequently counting them.

It would appear quicker to count a hundred, weigh them, and obtain the mean weight of one seed by shifting the decimal point. Actually, however, the time occupied in weighing with loose weights is much more than in weighing out seed to a fixed weight and then using the slide-rule. Ten grammes is a useful size of sample.

Lint Length.—Throughout the author's work on cotton this important feature has been determined by measurements made on the seed, and not by "pulling" the lint. The disadvantage of so doing is that the measurements do not coincide with the length of the pulled lint, as the grader, spinner, and trade, express it, and, moreover, seed-cotton must be at hand to measure.

The advantages far outweigh these disadvantages, as the method is far more accurate, and the conventional statement can always be obtained by the addition of a number which is constant for any given strain. Thus 33 millimetres "combed" length is 40 millimetres pulled" length. Many of the uncertainties which have crept into our beliefs about cotton may be traced to this handling of lint only, though, if it were practicable to adopt the method of measuring single fibres initiated by O'Neill, many of these could be eliminated.

The three methods are as follows:

(*a*) *Single-Fibre Measurements.*—Lint hairs whose ends are visible on the outside of a loose heap of lint are pulled out one by one, laid down with the wet finger on black paper, and measured with dividing compasses. (In all measurement work it is necessary to use dividers which are afterwards placed on a scale, since subjective error comes in if the scale is directly applied.) Provided that microscopic examination is included in the programme to insure that each fibre is unbroken, the result in good samples has a probable error for single fibres of about 10 per cent. O'Neill's original figures for Sea Island work out at about 8½ per cent. Thus the measurement of 100 fibres gives a probable error of 1 per cent., or 25 fibres 2 per cent. We shall see below that the same precision may be obtained by combing six seeds only as by measuring 25 fibres. There can be no doubt as to the relative ease of manipulation; single fibres are quite easy to handle in a good light, but any prolonged work with them strains the eyes severely, as the author knows only too well.

(*b*) *Pulling.*—The grader takes opposite sides of a lump of lint in the whole grip of each hand and draws them apart; he then grips the projecting fringe of one half between finger and thumb along a straight line and draws again; the distance from the grip to the end of the fringe gives the length, and the "hardness of the edge" gives the regularity of length. The details of pulling vary between different places, and between the graders of different kinds of cotton, some laying the pulled fibres repeatedly over one another, and then extracting a tuft from this parallelized group and placing it on the coat sleeve. This latter method is the more objectionable of the two. in that it extricates the longest fibres every time; but the same objection applies to both, namely, that the measurement is a measurement of the longest fibres and of the strongest fibres. This is unimportant to the skilled grader, who knows instinctively how to allow for it, but it leads to complete misunderstanding on the part of amateurs who attempt to copy him; and it should be remembered that any person who has spent less than ten years in the daily grading of cotton, and has not in addition been born with the instinct implanted in him, is an amateur at cotton-grading.

Further, it is not easy to measure the exact length of a pulled tuft, it having two vaguely defined ends instead of one, as the lint has which is combed *in situ* on the seed, and measurements are therefore just four times as incorrect.

(c) *Combed Seed-Cotton.*—The sample of seed-cotton to be measured is broken up into five, seven, or ten bundles, according to the accuracy desired, seven being the usual number. Each bundle is picked up in turn, pulled into two halves, and the first seed seen separating from the rest in the gap is picked out for combing; the choice of the seed is this way is nearly random, and if it is dependent on any property at all, it probably depends upon the twist of its lint, and not upon length.

Each seed is then combed with a small-tooth comb from the tip of the seed towards the butt, and outwards, at first lightly to disentangle the basal portion of the hairs, and then firmly, holding the seed and all the disentangled basal portions tightly between the finger and thumb of the left hand to prevent them from being torn apart. Finally a few strokes of the comb carry away any broken or detached fibres, and the seed is left with a halo of lint around it, chiefly at the basal portion. If the regularity of the lint is also under examination, more careful combing is employed to set each hair out along the radius of a circle with its centre in the seed.

The seed with its flat halo of lint is laid on a dark background, held down by the forefinger of the left hand resting on the seed; one leg of the dividers is then brought up against the butt of the seed, and the other is swung around and adjusted until it moves along the edge of the halo. Successive measurements made in this way on the same seed vary only 1 millimetre, so that the halo edge is obviously quite definite, even in poor cotton.

The mean of seven measurements has a probable error of less than 2 per cent., even in samples which consist of cotton damaged by premature opening, boll-worm, etc., and is thus better than twenty-five measurements made on single fibres.

The time occupied in the complete cycle of operations is ten minutes, or, when two persons are working together, about four minutes.

The reason for the superiority of the seed-combing method is obvious; it eliminates systematic variations of length between various parts of the seed, and measures the mean maximum length, leaving only the fluctuation from seed to seed to be wiped out by sampling. The application of statistical methods has been of the utmost service in this matter, as may well be realized on consulting works which quarrel with O'Neill's original method, because damping causes the fibre to stretch a millimetre, or for similar reasons. We are now able to prepare a numerical statement, of any desired degree of precision, in a reasonably short time and without eye-strain

But it should be observed that such a method is a research tool, or a means of exchanging ideas, and no more. While we amateurs are working out the lint length of a sample by ten minutes of effort, the grader will satisfy himself in as many seconds. It is also curious that the precise method should be the reverse of the convenient method, for the apparent length of cotton on the seed—even after combing—is most deceptive to the eye.

Examination of the Lint.—The chief characteristic examined in the ginned lint, apart from grading, is the strength, since length is preferably determined in the seed-cotton state. Before treating of strength-testing we may mention one or two minor features for which special methods have been found useful.

Diameter.—There is a more speedy way of determining diameter than by the micrometer eyepiece. If a camera lucida is set up, the magnification of its setting determined by drawing an object micrometer scale with it, and the diameter of fibres then drawn on a fresh sheet of paper, the mean diameter of a large number of fibres can be obtained very quickly. The fibres are mounted in a parallelized tuft, and observed in the middle of the tuft, or elsewhere as may be desired, and an **H**-shaped mark made on the drawing paper to define the two margins of each fibre, the cross-bar of the **H** tying together the two parallel margins in order to prevent confusion. The magnification should be so adjusted that each **H** is not less than 10 millimetres wide, and the mean width can then be taken with a millimetre scale, or by a simple form of instrument for totalling small lengths, which consists of a lever bearing a stylus and revolving a drum by means of a friction ratchet, or—where cheap labour is available—by cutting out each **H** from the paper, and placing them edge to edge in a row.

Weight.—The comparison of weight of equal lengths of fibre might be exceedingly useful, but until we can devise a machine which will count single lint hairs it must remain impracticable,* on account of the strain on the eyes, which is far worse than in isolating single fibres. When impact-testing of strength is employed, the two may be combined, and the fibre-counting necessitated by the one be utilized for the other.

Some seventy to a hundred fibres having been counted out, they are fixed across a gap in a piece of stout paper under slight tension by a drop of sealing-wax at either end. Twenty millimetres are then cut out of the centre by scissors, or by two safety razor blades mounted parallel in a brass holder. The bunch of 2-centimetre lengths thus obtained is bundled up and hung on the hook of a micro-

* See, however, note on p. 102.

balance, and the weight calculated to that of 1 centimetre of a single fibre.

The micro-balance used by the author was home-made, the torsion spring being fine capillary glass rod, mounted with sealing-wax at either end into a frame made of glass tube. The transverse lever was made of fine capillary glass tube, with a hook at one end to hold the fibres, and a counterpoising tail, the motion of which was observed in a mirror; a drop of sealing-wax united the lever to the torsion rod where they crossed, and the particular instrument employed was thus easily made to give 50 millimetres deflection for 1 milligramme, which was sufficiently sensitive for preliminary purposes.

The probable error of fibre-weight determinations would seem to be high, but the greater part of this is due to difficulties in sampling.

Strength.—The original work upon the breaking strain of single fibres is that of O'Neill, who rightly observes that: "Experimenters appear to have been deterred from manipulating with the individual hairs, on account of their smallness and lightness." O'Neill's paper has suffered from endless citation, but it was a very neat and accurate piece of work. The number of fibres he examined was not sufficient to give the general certainty to his figures which have since been attributed to them, only 363 fibres in all having been tested from seventeen different samples, representing about two weeks' steady work with the method he employed. But he published all his figures. From these we can work out the statistical significance, which is the same as Mr. Hughes and the author have found—namely, for length a probable error of about 10 per cent., and for breaking strain about 15 per cent.

O'Neill used a cylinder floating in water which could be withdrawn from a stopcock, thus increasing the strain on the fibre fastened to the top of the cylinder. Subsequent workers have modified his apparatus, notably Yves Henry; but the method remained slow—about fifteen minutes per test—and required the use of skilled labour throughout. Mr. F. Hughes made a great advance in the method by mounting the fibres across a hole in a piece of black paper with sealing-wax; this paper could then be hooked into the testing apparatus, the sides of the hole cut through, and the fibre was then free to be strained; we subsequently found that this device had been employed independently by other workers, but the merit of it from our point of view was that all the preparation could be put into the hands of a native lab-boy, and only the actual testing done by skilled labour, at the rate of about five minutes per test. Mr. Hughes further arranged that the load should be applied at a fairly constant velocity, by putting a fine-drawn tube in the stopcock outlet, and thus obviated one of the main objections to the old form of the appliance.

The immediate cause of this strength-testing work was our mild dissatisfaction with the spinning industry, who complained that Egyptian cotton was not so strong as it used to be, but could not produce sufficient figures to carry conviction to the growers and official bodies concerned, although many were prepared to believe it. It was felt, quite rightly, that some beginning must be made in keeping numerical records of some sort; and since it was obviously impossible to set up spinning tests in Egypt, data as to breaking strain of single fibres were better than nothing.

From Mr. Hughes' data it appeared that with proper sampling the probable error of single fibres was 15 per cent., so that a test made on fifty-six fibres would have a probable error of 2 per cent., or, in other words, that strength-testing could be made as accurate as length-testing without using an enormous number of fibres. Even at five minutes per fibre this meant an hour for each sample, and, as the author had one series requiring testing which alone consisted of sixty samples, he cast about for some method of speeding up the process.

Automatic Tester.—The outcome of some weeks of instrument-making in spare moments was a home-made machine which tested fibres one by one automatically, at the rate of one in twenty-five seconds, and single-fibre testing became practicable on a large scale. At the same time the instrument is—or, rather, was—a purely laboratory appliance, since it necessitated the native lab-boy's assistance to mount up the fibres on cards somewhat similar to those used by Mr. Hughes; these cards were loaded in a magazine, and the author's part in the testing consisted in aligning the magazine to the tester proper, cutting the cards and pulling over a switch. The magazine then swung in to the testing-points and placed a pair of half-cards upon them, with 10 mm. of fibre connecting them, moved away, and stopped. The tester then strained the fibre at a constant rate against a spring-balance which

PLATE XVI.—AUTOMATIC FIBRE-TESTER.

Instrument for testing the breaking-strain of single fibres, which are mounted on cards, loaded upon the magazine (left), which passes them one by one to the tester (centre), the strain being recorded on the drum (right). (See p. 190.)

recorded the strain on a drum. When the fibre gave way the balance returned to rest, the drum moved on to receive a fresh trace, the broken fibre with its cards was thrown off, and, the clutch of the magazine being pulled out at the same time, the cycle of operations started again.

The set of cards in the magazine having been dealt with, the paper was removed from the drum, and a fresh magazine inserted, with a fresh batch of fibres. The mean breaking strain was obtained quickly by adding up the total deflections of the balance as marked on the drum with a map-measurer, the whole operation with a magazine of twenty fibres taking eight minutes, most of which was simply spent in watching the machine do the work (Pl. XVI.).

Impact Testing.—On leaving Egypt the author lost this automatic tester, because some parts of it had been made with Government material. The task of reconstructing it was rather formidable, and a timely suggestion was derived from a paper by Mr. J. H. Lester, in which he points out the value of ballistic testing of yarn. The chief feature, from the author's view-point, is that bunches of yarns or fibres can be tested together. If several fibres are slowly strained to determine the breaking strain, the rupture of the first one throws its share of the load on the others, which therefore yield in rapid succession, and the result is meaningless.

In impact testing, on the other hand, the force measured is kinetic, the breaking of each fibre subtracting a definite amount of kinetic energy.

The simple home-made form of the implement devised

by the author consists of a pendulum, a catch to hold it at a definite altitude, and a smoked plate on which the pendulum traces its swing with a delicate bristle stylus. At the lowest point of the swing the apex of the pendulum meets the end of a slot in a piece of tough paper and carries it along with it. In so doing a bunch of fibres is broken, these fibres having been mounted across a gap 10 millimetres wide in the after-portion of the paper, and left to bear the shock of the impact by cutting the sides of the gap as before, the portion of paper remaining behind the gap being held firmly by a peg. The pendulum swings up to a certain point when it has only the inertia of the slotted piece of paper to overcome; when it also has to fracture a fibre or bunch of fibres, it swings up to a less extent; the difference is measured from the smoked trace, and gives by calculation the number of gramme-centimetres of energy expended in breaking the fibres, or (on dividing by the number of fibres) the resistance to impact of a single fibre.

The method has great advantages, but requires very careful sampling, if the full advantage of speed is to be secured, and it also necessitates the counting of single fibres. It will probably be possible to introduce a machine which will obviate both these disadvantages, and make such testing a practicable piece of routine. The time occupied in making the tests with the ordinary machine works out at ten minutes for each complete cycle of operations on each bunch of ten to twenty fibres, five of these being tested for each sample. The probable error of determinations thus made is very low, since a bunch of sixteen fibres has only a quarter of the probable error of a single fibre. The whole difficulty lies in sampling, and the results obtained with the impact tester will serve to illustrate the points in general.

Sampling.—It might appear easy to obtain a uniform lot of fibres by repeated drawing of the lint, overlaying tufts on one another, and drawing again from these. In point of fact the only reliable way is to pick the hairs one by one at random from those projecting out of a loose lump of lint, as O'Neill did. Again, it should be noticed, the amateur and the grader obtain the best results from exactly opposite methods

Illustrating the difficulty of sampling from numerical obtained with the impact tester, and leaving the numbers in their arbitrary scale, taken direct from the notebook, we obtain such results as the following:

Small tuft of about 500 fibres taken from a square millimetre of the butt of a single seed, and tested in bunches of 4, 5, 7, 10, 12, 14, 16, and 18 fibres respectively, worked out at the following strengths per fibre: 1·25, 1·60, 1·85, 2·00, 1·40, 1·85, 1·70, and 1·60. The variation is slight, though the absolute strength is very low.

Another seed of the same sample was tested from six points round it, two on either side—one from the short hairs near the tip, and one from the butt; twelve hairs in each sample; strength per fibre was—Tip 2·5, butt 4·2, left side 5·4 and 6·4, right side 3·1 and 3·1. The variation is increased, probably because the nutrition varies according to the proximity of the particular hairs to the vascular bundles which supply food-substances.

Taking next a small tuft picked at random from a good sample of pure-strain ginned lint, and testing successive bunches from it, we find a moderate amount of variation due to the previous causes: Six fibres, 2·8, 3·3, 3·5, 3·7, 4·0; eight fibres, 3·5, 2·2; ten fibres, 3·2, 4·1; twelve fibres, 4·2, 5·9; fourteen fibres, 3·6, 3·4, 3·4, 3·3. It will be noticed with sufficient clearness for our present purpose how the probable error is decreasing as larger bunches of fibres are taken, and the fibre to fibre variation is eliminated thereby

If now, instead of taking a small tuft from one part of a good ginned sample, we take tufts from seed-cotton, such as were used in one of the series hereafter to be described, we find that it is impossible to sample effectively. The strength of the lint in the 1913 series of dated bolls was—owing to circumstances—necessarily determined from unginned cotton. In order to obviate the disadvantage of losing the mixing action of the gin, the following routine was practised. Each sample was broken into five lots, one for each impact test; each lot contained usually from fifty to 500 seeds, and ten tufts of lint were drawn from each lot, each tuft taking some fibre from not less than three seeds. These ten tufts were rolled up together, then pulled straight, and a wide layer drawn out as in grading for regularity; this layer was then drawn down right and left until a countable number of fibres remained, and these were tested. The figures obtained with all these precautions are given subsequently, and it will be seen that, although they give five-day means with a probable error of 5 per cent., which is good enough for our purpose, the bunches so carefully extracted in such a way as to include fibres from all parts of each of the five lots are very often extremely wild; the probable error for the average of a single bunch is 24 per cent. Had it been practicable to effect the work on ginned lint, this figure would have been very much lower; but it is quite useful to have a detailed example of these sampling difficulties, for the routine method just described was most strictly followed in obtaining every bunch tested in the Daily Picking Series of 1912.

Grading.—The only grading data included in the account of the dated samples are those for strength. The determinations were made by Mr. Harold C. Thomas, of the National Bank of Egypt, Alexandria, and they are a striking instance of the accuracy to which the grader's hands can attain; the samples graded were about 8 grammes in weight, and were given to Mr. Thomas in irregular sequence, in three separate batches, marked with dummy reference numbers which bore no relation to their actual daily sequence; Mr. Thomas knew only that he was grading these dated samples from a familiar pure strain. In spite of these precautions against subjectivity, it will be seen that his hands assigned sample after sample to what was obviously its correct place in relation to its neighbours, and that there are only one or two wild points in his strength curve.

Such results emphasize the futility of attempting to introduce so-called "scientific methods" into the ordinary commercial practice; grading by hand has its limitations, and so have the scientific methods; each has its proper function, and the results of each are of interest to the other.

APPENDIX II

TABLES OF STATISTICAL DATA

These tables embody the results discussed in Chapter IV., and they show the properties of samples of cotton formed on successive days of the season. The same "pure strain" of cotton was employed in both the Dated Flowers Experiment of 1912 and the Daily Picking Experiment of 1913, so that any differences between different samples are solely due to the action of the environment on the plants, and not to any differences of inherited constitution.

The differences between various samples as numerically expressed in these tables embody in addition certain unavoidable experimental errors, so that two samples might be identical and yet not give exactly the same numerical results. These errors have been reduced as far as was practicable in the execution of the work, and have been further obliterated by working out "five-day means."

Tables I. and III. present the actual experimental figures for each lint hair, seed, or sample, examined in the two experiments. Tables II. and IV. summarize the average results for each five days. Thus the figures given in these tables on August 14 are the means of the corresponding figures for August 12, 13, 14, 15, and 16, in Tables I. and III.; while the figures for August 15 are the corresponding means for August 13, 14, 15, 16, and 17. The five-day means thus obtained are the data plotted in the curves of Figs. 14 and 15.

TABLE I.

DATED FLOWERS, 1912 (ACTUAL EXPERIMENTAL DATA).

Flower opened.	Lint Length (Combed on Seed).		Breaking Strain of Lint Hairs.		Seed-Cotton.		Seed Weight in Grms.	Lint Weight in Grms.	Grader's "Strength."
	Each Seed in Millimetres.	Mean.	Each Hair in Grammes (about).	Mean.	Weighed in Grms.	Out-turn Per Cent.			
July 7	27, 27, 31, 31, 32, 33, 33, 34	31·0	3, 5, 5, 6, 6, 7, 8 (3), 9, 10, 10, 11 (3)	8·0	19·58	26·2	0·0990	0·0351	W
8	27, 28, 28, 30, 30, 30, 31, 33	29·5	1, 3, 3, 4, 4, 5, 6, 6, 7, 7, 8, 9, 9, 10 (3), 11 (3), 12, 12	7·6	21·75	28·3	0·1060	0·0417	M
9	26, 27, 28, 28, 30, 30, 33, 33	29·4	2, 3 (3), 4 (3), 5 (3), 6, 7, 7, 8, 8, 10, 12, 13	6·0	22·13	26·8	0·0980	0·0359	S
10	24, 28, 28, 30, 31, 31, 33, 34	29·9	2, 2, 3, 3, 4, 5, 5, 6, 6, 7 (3), 8 (3), 10, 12, 13	6·8	19·79	25·8	0·1080	0·0376	S
11	28, 30, 30, 31, 31, 32, 33, 35	31·2	4, 5 (3), 6, 7 (3), 8, 8, 10, 10, 11, 11, 12, 12, 13, 14	8·4	22 51	27·5	0·1090	0·0412	SS
12	28, 29, 29, 29, 30, 31, 32, 33	30·1	3, 3, 4, 5, 7, 8 (4), 9, 9, 10 (3), 12, 12, 13, 13	8·3	25·86	26·5	0·1080	0·0390	SS
13	28, 29, 30, 30, 30, 31, 31, 32	30·1	2, 2, 3, 3, 4, 4, 5, 6, 6, 7, 7, 8, 8, 9, 9, 10, 10	6·0	26·13	29·6	0·1080	0·0455	SSS
14	27, 28, 29, 29, 30, 30, 31, 33	29·6	3, 3, 4 (4), 6, 6, 8, 9, 11, 11, 12, 13, 13, 14, 14	8·1	21·18	28·8	0·1070	0·0433	SSS
15	27, 28, 28, 29, 29, 30, 31, 31	29·1	3, 4 (4), 5, 5, 6, 7, 8 (3), 10, 10, 11, 11, 14, 14, 15	8·1	23·63	30·0	0·1110	0·0475	SSS
16	Ginned, by mistake	—	1, 2, 2, 3, 3, 4, 4, 5, 6, 6, 7, 7, 8, 9, 9, 11, 12, 13	6·3	22·90	29·7	0·1050	0·0444	SSS
17	27, 28, 28, 29, 29, 31, 32, 32	29·5	3, 3, 5 (4), 6, 8, 8, 9, 9, 10 (3), 11, 12, 12, 14, 14	7·6	22·08	27·7	0·1080	0·0415	SSS
18	25, 29, 29, 30, 31, 31, 32, 32	29·9	1, 2, 6, 9, 9, 10, 12, 12, 13 (3), 15, 15, 16, 17	10·9	23·11	29·5	0·1100	0·0461	SSSS
19	26, 27, 28, 29, 29, 30, 30, 31	28·7	3, 6, 8 (4), 9 (3), 10 (4), 12, 12, 13, 14, 15, 16, 17	10·0	21·85	29·5	0·1100	0·0461	SSSS
20	26, 28, 29, 30, 30, 30, 31, 32	29·5	2, 3, 4, 5, 7, 7, 8, 8, 9, 10, 11, 11, 13, 14, 17, 20	9·7	18·71	30·8	0·1080	0·0480	SS

21	27, 28, 29, 29, 30, 30, 30, 31	29·2	2, 3, 4, 6, 7, 8, 8, 10, 11, 12, 13 (3), 14, 14, 17, 17, 22	10·5	19·45	28·9	0·1070	0·0436	S
22	26, 26, 27, 28, 29, 30, 33, 34	29·1	3, 4, 4, 5 (3), 6, 6, 7, 7, 10 (3), 11, 12, 12, 14, 14, 15, 15	8·7	18·87	30·1	0·1050	0·0452	SS
23	26, 26, 27, 28, 28, 29, 30, 31	28·1	1, 4, 6, 7, 7, 8, 9, 9, 10 (3), 11, 12, 12, 13, 13, 14 (3)	10·0	10·11	28·9	0·1060	0·0431	S
24	27, 29, 30, 31, 32	29·8	4, 6, 7, 7, 9, 9, 10 (3), 11, 12, 13, 13, 16, 18, 22	10·5	21·11	32·8	0·1110	0·0541	SSS
25	28, 29, 30, 30, 31	29·6	1 (3), 2, 2, 4, 5 (3), 6, 7 (3), 8, 8, 9, 12, 15	6·5	19·77	30·0	0·1000	0·0428	SSS
26	30, 30, 30, 31, 31	30·4	3, 4, 4, 5, 6, 6, 7, 7, 8 (3), 9, 9, 11, 12, 14 (3)	8·6	18·79	29·1	0·0960	0·0393	SS
27	25, 26, 28, 28, 29	27·2	2, 3, 4, 4, 5, 5, 6 (3), 7, 8, 8, 9, 9, 11, 11, 13, 14, 19	7·8	26·69	27·8	0·1010	0·0388	SS
28	27, 28, 28, 28, 30	28·2	2, 7, 9, 10 (3), 11, 11, 12 (3), 13 (3), 14, 16, 17, 17, 18, 20, 21	12·4	24·78	28·8	0·1030	0·0417	SSS
29	25, 26, 26, 27, 28	26·4	1, 4, 4, 5, 6, 6, 7, 8, 8, 8, 9, 10 (3), 11, 12, 13 (3), 14, 15, 16, 19	11·1	14·31	29·0	0·0960	0·0392	SS
30	25, 27, 28, 30, 30	28·0	1, 3, 4, 6, 6, 9, 9, 10, 10, 11, 11, 13 (3), 14, 18, 22, 22	11·6	21·04	31·3	0·1070	0·0488	SS
31	23, 26, 31, 31, 34	29·0	3 (3), 5, 6, 6, 7, 8, 9, 9, 12, 12, 13, 14, 14, 16, 17	9·5	22·94	29·2	0·1070	0·0448	SS
Aug. 1	27, 28, 29, 29, 30	28·6	2, 2, 3, 3, 4, 4, 5, 5, 6, 7, 8, 10 (3), 11, 13, 14, 17, 19	9·0	25·39	29·0	0·1020	0·0415	W
2	28, 28, 29, 30, 31, 32, 35	30·4	4, 5 (3), 6, 7, 8, 8, 9, 10, 10, 11 (4), 12, 12, 13, 18	9·5	19·20	31·0	0·0950	0·0426	SSS
3	27, 28, 29, 31, 30, 30, 33	29·7	2, 3 (3), 4, 4, 5 (4), 6, 10, 11 (3), 12, 15	6·5	16·53	29·7	0·0970	0·0411	SSS
4	25, 26, 27, 27, 29, 30, 32	28·0	1, 6, 6, 8, 8, 9, 11, 12, 12, 13, 15, 18, 24	12·5	15·54	29·7	0·0950	0·0403	M
5	29, 29, 29, 31, 31, 32, 32	30·4	1, 2, 3, 4, 6, 6, 7, 8, 9 (3), 10, 10, 11, 12, 12, 21	8·4	—	—	0·1010	—	M
6	27, 30, 31, 31, 31, 32, 32	30·6	1, 4 (3), 5, 5, 6, 6, 7 (3), 8 (4), 11, 12, 13, 13, 15	7·5	15·76	28·6	0·0860	0·0398	M
7	28, 30, 30, 30, 31, 32, 32	30·4	2, 3 (3), 4, 5, 5, 6 (3), 7, 7, 9, 10, 10, 11, 13, 14, 15, 17	8·7	21·47	31·6	0·0990	0·0457	SSS
8	29, 29, 29, 30, 31, 31, 32	30·1	—	9·3	23·91	27·9	0·1060	0·0410	SSS
9	26, 30, 30, 30, 31, 32, 34	30·4	4, 5, 7, 9, 9, 10, 11, 12, 13, 15 (4), 18, 23	12·5	18·41	29·9	0·0980	0·0418	S
10	30, 31, 32, 32, 32, 33, 33	31·8	1, 2, 2, 3, 3, 4 (4), 5, 5, 6 (3), 9, 10, 14, 18, 21	7·4	20·60	26·5	0·0980	0·0354	SSSS

TABLE I.—*Continued.*

DATED FLOWERS, 1912 (ACTUAL EXPERIMENTAL DATA).

Flower opened.	Lint Length (Combed on Seed).		Breaking Strain of Lint Hairs.		Seed-Cotton.		Seed Weight in Grms.	Lint Weight in Grms.	Grader's "Strength."
	Each Seed in Millimetres.	Mean.	Each Hair in Grammes (about).	Mean.	Weighed in Grms.	Out-turn Per Cent.			
Aug. 11	29, 30, 30, 30, 30, 31, 33	30·4	1, 3, 3, 4, 4, 8, 10, 10, 11, 13 (3), 16, 20	9·4	20·00	30·3	0·1000	0·0434	S
12	28, 30, 30, 31, 33, 34, 34	31·4	1, 1, 2, 4, 4, 5, 5, 6, 7 (3), 8, 9, 9, 11, 11, 12, 14, 21	8·0	17·05	28·3	0·0860	0·0340	SSS
13	31, 31, 31, 32, 32, 32, 33	31·7	1, 2, 2, 3, 4, 5, 5, 7 (3), 8, 8, 9, 9, 10, 11, 12, 13, 18	7·3	19·55	31·4	0·0870	0·0398	SS
14	25, 26, 26, 28, 29, 32, 32	28·3	1, 3, 3, 4, 6 (3), 7, 7, 8, 8, 9, 10 (3), 11, 11, 13, 14	7·5	16·98	29·6	0·0930	0·0392	SSS
15	27, 28, 30, 31, 31, 32, 33	30·3	2 (4), 3 (4), 4, 7, 8 (3), 9 (3), 11, 13	6·0	10·63	31·3	0·0920	0·0418	S
16	28, 28, 29, 29, 32, 32	30·0	2, 3, 4, 4, 5, 6, 7, 8, 8, 11, 11, 12, 13, 13, 14, 19	8·6	20·66	30·0	0·0930	0·0398	S
17	28, 29, 30, 30, 30, 31, 33	30·1	—	9·2	17·58	29·6	0·0990	0·0415	SS
18	26, 29, 29, 30, 30, 32, 32	29·7	3, 5, 6, 8, 10 (3), 15	8·0	13·86	30·7	0·0880	0·0389	W
19	26, 28, 29, 31, 31, 32, 32	29·9	—	4·5	13·28	30·5	0·0810	0·0355	VW
20	28, 33, 32, 32, 32, 32, 35	31·9	4, 5, 5, 7, 7, 8, 9 (4), 10 (4), 11, 12, 12, 16	8·8	24·32	31·6	0·6860	0·0396	MS
21	28, 30, 30, 31, 31, 31, 31	30·3	—	6·5	16·66	30·7	0·0880	0·0388	S
22	29, 30, 31, 31, 31, 33, 33	31·1	—	8·0	15·35	31·0	0·0790	0·0355	SSS
23	25, 26, 27, 29, 29, 30, 31	28·1	—	8·0	7·56	32·3	0·0840	0·0401	W
24	24, 27, 27, 28, 28, 29, 30	27·6	1, 1, 2 (3), 3, 4, 4, 5, 6 (3), 7, 8, 9, 9, 10, 11, 13	6·0	17·04	29·5	0·0830	0·0347	VW
25	28, 28, 29, 29, 30, 31, 32	29·6	2, 3 (4), 4, 4, 5, 7, 7, 8, 8, 10, 10, 12, 13, 14, 14, 15	7·0	12·71	30·9	0·0780	0·0349	MS
26	24, 28, 28, 29, 29, 30, 32	28·6	—	7·5	14·39	31·5	0·0760	0·0349	W
27	30, 31, 31, 31, 32, 32, 33	31·4	—	7·5	16·08	30·5	0·0780	0·0342	M

28	26, 28, 28, 28, 28, 29, 32	28·4	4, 4, 8 (5), 9 (4), 10 (4), 12, 13, 13, 15	9·0	6·95	28·9	0·0720	0·0342	VW
29	25, 28, 28, 29, 31, 32, 32	29·6	—	11·8	15·59	29·6	0·0880	0·0370	S
30	26, 28, 28, 30, 30, 31, 32	29·1	—	5·0	15·40	31·4	0·0810	0·0370	S
31	25, 29, 29, 30, 30, 30, 31	29·1	4 (4), 5, 6 (4), 8, 8, 9, 10, 10, 11, 13, 14	7·5	9·77	31·0	0·0780	0·0350	W
Sept.									
1	25, 25, 25, 28, 29, 30, 31	27·6	—	6·8	18·16	30·2	0·0770	0·0332	MS
2	—	—	—	5·0	9·19	31·5	0·0710	0·0325	SS
3	—	—	—	4·5	3·72	32·3	0·0660	0·0315	M
4	—	—	—	3·7	5·97	29·1	0·0550	—	VW
5	—	—	—	2·8	3·81	27·2	—	—	MS
6	—	—	—	1·7	—	—	—	—	SS

NOTES TO TABLE I.

The breaking strains for each hair tested are given. Where more than two hairs gave the same breaking strain, the number so doing is given in brackets after the breaking-strain figure.

The breaking strain is given in "grammes (about)." Actually the unit was slightly under 1 gramme, so that the strongest fibre broke under a load of about 22 grammes, and not of 24 grammes as denoted. The difference is almost negligible. The figures for each hair are omitted in certain samples, owing to the original record on the drum of the fibre-tester having been lost. The mean value had alone been preserved in a duplicate entry.

The column "Seed-Cotton" gives the actual weight of sample handled, as well as its out-turn. "Seed Weight" gives the average weight of a single seed in each sample, and "Lint Weight" gives the average weight of lint on a single seed as computed from the ginning out-turn.

The column headed "Grader's 'Strength'" gives the strengths assigned by the grader to each sample, using the following scale:

Very weak	VW	Strong	S
Weak	W	Stronger	SS
Moderate	M	Very strong	SSS
Medium strong	MS	Superlatively strong	SSSS

TABLE II.

DATED FLOWERS, 1912
(DATA SMOOTHED TO FIVE-DAY MEANS).

Date.	Lint.				Out-turn per Cent.	Seed Weight in Grammes.
	Length in Millimetres.	Breaking Strain in Grammes.	Weight per Seed in Grammes.	Weight Regression.		
July 7	—	—	—	—	—	—
,, 8	—	—	—	—	—	—
,, 9	30·25	7·2	0·0385	0·0365	27·0	0·104
,, 10	30·15	7·6	0·0390	0·0375	26·8	0·106
,, 11	30·15	7·7	0·0398	0·0372	27·1	0·106
,, 12	30·20	7·7	0·0412	0·0372	27·7	0·107
,, 13	29·10	7·8	0·0432	0·0400	28·6	0·108
,, 14	29·60	7·6	0·0438	0·0410	28·8	0·107
,, 15	29·50	7·6	0·0445	0·0415	29·2	0·108
,, 16	29·50	8·2	0·0442	0·0420	29·2	0·108
,, 17	29·40	8·4	0·0450	0·0425	29·2	0·108
,, 18	29·50	9·0	0·0451	0·0430	29·5	0·108
,, 19	29·30	9·7	0·0451	0·0428	29·2	0·108
,, 20	29·35	10·0	0·0451	0·0440	29·8	0·108
,, 21	29·90	9·9	0·0451	0·0440	29·6	0·107
,, 22	29·25	9·9	0·0475	0·0452	30·3	0·107
,, 23	29·25	9·4	0·0451	0·0448	30·2	0·105
,, 24	29·45	9·0	0·0450	0·0442	30·2	0·104
,, 25	29·20	8·6	0·0435	0·0435	29·6	0·102
,, 26	29·30	9·0	0·0438	0·0430	29·6	0·101
,, 27	28·70	9·3	0·0400	0·0430	29·0	0·099
,, 28	28·40	10·3	0·0412	0·0440	29·2	0·100
,, 29	28·20	10·4	0·0422	0·0445	29·0	0·102
,, 30	28·30	10·7	0·0455	0·0445	29·4	0·103
,, 31	28·50	10·0	0·0442	0·0448	29·8	0·102
Aug. 1	29·15	9·2	0·0435	0·0442	30·0	0·101
,, 2	29·10	9·2	0·0422	0·0440	29·8	0·100
,, 3	29·50	9·2	0·0415	0·0438	29·7	0·098
,, 4	29·80	8·8	0·0410	0·0430	29·7	0·095
,, 5	29·90	8·6	0·0410	0·0428	29·6	0·096
,, 6	29·85	9·1	0·0410	0·0430	29·2	0·098
,, 7	30·30	9·4	0·0415	0·0420	29·1	0·098
,, 8	30·60	9·4	0·0408	0·0410	28·6	0·097
,, 9	30·65	9·6	0·0412	0·0402	28·9	0·100
,, 10	30·85	9·3	0·0395	0·0400	28·6	0·097

TABLE II.—*Continued*

DATED FLOWERS, 1912

(DATA SMOOTHED TO FIVE-DAY MEANS).

Date.	Lint.				Out-turn per Cent.	Seed Weight in Grammes.
	Length in Millimetres.	Breaking Strain in Grammes.	Weight per Seed in Grammes.	Weight Regression.		
Aug. 11	31·10	9·0	0·0392	0·0400	29·2	0·094
,, 12	30·80	8·0	0·0382	0·0405	29·3	0·092
,, 13	30·40	7·6	0·0395	0·0425	30·2	0·091
,, 14	30·30	7·4	0·0392	0·0430	30·1	0·090
,, 15	30·10	7·8	0·0408	0·0440	30·4	0·092
,, 16	29·95	7·9	0·0405	0·0435	30·1	0·093
,, 17	29·95	7·4	0·0402	0·0438	30·4	0·090
,, 18	30·30	7·8	0·0385	0·0432	30·3	0·089
,, 19	30·45	7·4	0·0383	0·0438	30·5	0·089
,, 20	30·60	7·3	0·0380	0·0440	30·9	0·085
,, 21	30·30	7·2	0·0378	0·0450	31·1	0·085
,, 22	29·85	7·4	0·0375	0·0452	31·0	0·085
,, 23	29·30	7·2	0·0372	0·0480	30·9	0·082
,, 24	29·00	7·2	0·0358	0·0520	30·9	0·080
,, 25	29·00	7·2	0·0358	0·0525	31·1	0·080
,, 26	29·10	7·3	0·0348	0·0525	30·3	0·077
,, 27	29·50	8·5	0·0348	0·0515	30·3	0·078
,, 28	29·35	8·4	0·0350	0·0490	30·3	0·079
,, 29	29·40	8·4	0·0360	0·0470	30·2	0·079
,, 30	28·70	7·7	0·0355	0·0480	30·2	0·080
,, 31	—	7·2	0·0348	—	30·7	0·079
Sept. 1	—	6·0	0·0338	—	31·2	0·075
,, 2	—	5·0	0·0315	—	31·0	0·070
,, 3	—	4·4	0·0305	—	—	—
,, 4	—	3·7	—	—	—	—
,, 5	—	2·8	—	—	—	—
,, 6	—	2·0	—	—	—	—

NOTES TO TABLE II.

July 7 and 8 are necessarily blank because there are no precedent data with which to form a five-day mean for them.

The columns for "Length," "Breaking Strain," "Lint Weight per Seed," "Out-turn," and "Seed Weight," are computed to five-day means directly from the daily data in Table I.

The column headed "Lint Weight Regression" is additional. It represents the weight of lint which the seed would have borne each day *if* the seed had been constantly 0·100 gramme in weight, and *if* the lint had been constantly 30 millimetres long.

TABLE III.

DAILY PICKINGS, 1913 (ACTUAL EXPERIMENTAL DATA).

Date of Picking.	Number of Bolls picked per Plant.	Length of Lint in Millimetres by combing Seed-Cotton.		Strength of Lint (in Arbitrary Scale) by Impact Testing of Five Bundles of Hairs.		
		Each Seed combed.	Mean.	Each Bundle tested (Number in Bundle).	Total tested.	Mean.
Aug.						
12	0·008	—	—	1·61 (13), 1·71 (14), 3·50 (16), 5·91 (12), 8·00 (11)	64	4·03
13	0·008	—	—	1·45 (9), 2·41 (12), 4·18 (11), 8·60 (10), 6·57 (14)	57	4·65
14	0·015	—	—	2·64 (14), 5·43 (16), 7·06 (13), 6·48 (19), 4·50 (27)	89	5·25
15	0·061	—	—	3·78 (9), 3·78 (9), 5·81 (11), 5·50 (16), 7·50 (14)	59	5·39
16	0·054	—	—	2·00 (10), 2·33 (12), 2·26 (15), 2·50 (12), 2·87 (16)	65	2·60
17	0·031	—	—	1·13 (16), 3·00 (14), 3·57 (14), 4·83 (12), 3·89 (18)	74	3·24
18	0·092	—	—	1·55 (11), 1·59 (17), 3·67 (12), 2·61 (18), 3·47 (15)	73	2·57
19	0·107	—	—	2·44 (18), 5·22 (9), 4·14 (14), 8·46 (15), 5·30 (24)	80	5·07
20	0·107	—	—	2·00 (11), 1·86 (14), 2·45 (22), 2·72 (22), 4·66 (18)	87	2·78
21	0·069	—	—	2·88 (17), 3·29 (17), 3·36 (19), 3·95 (20), 3·76 (25)	98	3·47
22	0·038	—	—	3·33 (9), 2·46 (15), 2·93 (15), 3·22 (18), 3·31 (23)	80	3·05
23	0·077	—	—	2·00 (6), 1·80 (11), 4·66 (6), 6·55 (9), 4·20 (16)	48	3·21
24	0·200	—	—	3·58 (12), 3·61 (13), 3·26 (15), 3·94 (15), 4·61 (13)	68	3·80
25	0·115	—	—	3·60 (13), 3·50 (14), 5·00 (13), 6·78 (14), 4·40 (22)	76	4·65
26	0·031	—	—	3·86 (14), 5·00 (12), 6·80 (10), 6·07 (13), 7·07 (13)	62	5·73
27	0·207	—	—	1·00 (16), 1·77 (22), 4·29 (14), 3·79 (19), 5·35 (17)	88	3·20
28	0·115	—	—	5·00 (13), 5·61 (13), 3·70 (14), 5·85 (20), 4·90 (21)	81	4·95
29	0·107	—	—	1·70 (10), 7·00 (7), 5·42 (12), 4·71 (14), 4·12 (16)	59	4·52
30	0·177	—	—	1·14 (7), 2·89 (9), 2·61 (13), 5·34 (9), 6·11 (16)	54	3·79
31	0·215	—	—	2·33 (12), 5·08 (12), 6·50 (12), 5·73 (15), 4·50 (28)	79	4·87

Sept.						
1	0·200	32, 32, 33, 34, 36, 36, 37	34·3	7·25 (12), 6·63 (14), 6·50 (16), 5·90 (19), 6·89 (17)	78	6·61
2	0·185	32, 33, 33, 33, 36, 37, 38	34·6	2·63 (11), 5·81 (10), 4·61 (13), 4·81 (16), 5·76 (17)	67	4·76
3	0·215	29, 33, 33, 34, 35, 37, 37	34·0	1·40 (10), 2·75 (8), 2·31 (13), 2·89 (18), 4·68 (22)	71	2·99
4	0·253	35, 35, 35, 35, 36, 36, 36	35·4	2·16 (12), 2·53 (13), 2·80 (15), 2·00 (29), 6·00 (16)	85	3·05
5	0·130	29, 30, 32, 32, 34, 35, 35	33·9	6·25 (8), 5·30 (13), 5·50 (16), 6·06 (16), 6·42 (19)	72	5·91
6	0·338	29, 31, 33, 34, 34, 35, 35	33·0	1·45 (18), 3·73 (15), 3·73 (15), 4·76 (13), 4·89 (18)	79	3·67
7	0·138	32, 33, 34, 34, 35, 36, 38	34·6	1·66 (15), 1·58 (19), 2·50 (14), 3·12 (16), 5·25 (16)	80	2·81
8	0·469	28, 32, 35, 35, 35, 36, 36	33·9	2·74 (12), 3·00 (9), 2·60 (13), 1·88 (15), 4·24 (22)	71	2·96
9	0·315	31, 32, 34, 36, 36, 36, 36	34·4	2·82 (12), 2·40 (10), 1·84 (11), 1·18 (17), 5·84 (17)	67	2·88
10	0·215	30, 31, 33, 34, 34, 35, 37	33·4	4·26 (12), 5·29 (12), 3·24 (7), 6·35 (8), 5·52 (16)	82	5·00
11	0·261	32, 33, 34, 34, 35, 35, 36	34·1	1·69 (13), 2·53 (17), 5·76 (17), 6·00 (19), 5·65 (23)	89	4·45
12	0·146	31, 32, 33, 33, 33, 34, 36	33·1	1·08 (12), 1·07 (13), 2·00 (16), 5·19 (21), 4·10 (30)	92	3·92
13	0·069	28, 29, 31, 31, 34, 37, 38	32·6	2·61 (13), 6·55 (9), 4·80 (15), 5·06 (15), 4·30 (20)	72	4·60
14	0·515	29, 31, 32, 35, 35, 35, 37	33·4	1·25 (12), 1·33 (18), 2·75 (20), 4·61 (18), 5·95 (19)	87	3·25
15	0·491	30, 31, 31, 35, 35, 36, 37	33·6	2·56 (9), 3·00 (14), 4·33 (15), 4·10 (18), 6·67 (18)	74	4·25
16	0·484	32, 32, 32, 33, 33, 35, 35	33·1	1·37 (16), 2·21 (14), 3·70 (10), 3·64 (14), 4·93 (15)	69	3·14
17	0·491	33, 33, 34, 34, 35, 36, 37	34·4	3·18 (11), 5·20 (10), 5·06 (16), 6·50 (14), 6·52 (15)	66	5·35
18	0·177	31, 33, 34, 34, 34, 35, 35	33·7	1·30 (10), 2·16 (12), 5·21 (14), 5·54 (11), 5·85 (20)	67	4·17
19	0·276	31, 31, 33, 34, 34, 35, 36	33·4	0·83 (17), 1·07 (13), 2·18 (17), 2·38 (13), 4·53 (15)	75	2·19
20	0·591	33, 35, 35, 35, 35, 35, 36	34·9	1·78 (18), 3·90 (10), 4·46 (13), 4·61 (13), 5·18 (16)	70	3·94
21	0·376	32, 32, 34, 34, 35, 35, 36	34·0	2·00 (15), 2·80 (21), 2·84 (13), 3·16 (24), 3·92 (12)	85	2·93
22	0·277	31, 32, 32, 34, 34, 35, 37	33·6	0·91 (11), 1·00 (10), 3·66 (12), 4·11 (19), 6·00 (19)	71	3·37
23	0·353	32, 33, 33, 34, 35, 35, 36	34·1	2·31 (16), 3·87 (16), 5·61 (18), 6·10 (20), 6·80 (15)	85	4·97
24	0·599	31, 32, 33, 35, 35, 36, 37	34·1	2·00 (10), 5·00 (14), 5·90 (11), 6·50 (10), 6·70 (16)	61	5·29
25	0·246	30, 31, 32, 34, 37, 37, 37	34·0	2·00 (16), 2·78 (9), 2·79 (12), 5·31 (19), 7·38 (8)	64	4·05
26	0·323	32, 32, 33, 34, 35, 35, 38	34·1	4·07 (14), 4·30 (13), 4·30 (13), 4·57 (14), 4·90 (11)	65	4·42
27	0·376	31, 33, 33, 33, 35, 35, 36	33·7	1·89 (9), 4·33 (15), 4·73 (15), 5·26 (15), 6·16 (12)	66	4·55
28	0·276	32, 33, 33, 33, 35, 36, 37	34·1	2·00 (12), 2·07 (13), 3·60 (20), 4·90 (11), 11·00 (11)	67	4·58
29	0·315	27, 32, 33, 33, 34, 34, 36	32·7	4·08 (27), 4·61 (13), 5·62 (16), 5·93 (15), 6·50 (20)	91	5·31
30	0·161	29, 30, 30, 31, 33, 36, 36	32·1	1·23 (13), 1·66 (9), 1·85 (13), 1·88 (8), 1·92 (26)	69	1·73

TABLE III.—*Continued.*

DAILY PICKINGS, 1913 (ACTUAL EXPERIMENTAL DATA).

Date of Picking.	Number of Bolls picked per Plant.	Length of Lint in Millimetres by combing Seed-Cotton.		Strength of Lint (in Arbitrary Scale) by Impact Testing of Five Bundles of Hairs.		
		Each Seed combed.	Mean.	Each Bundle tested (Number in Bundle).	Total tested.	Mean.
Oct.						
1	0·261	31, 31, 32, 33, 33, 34, 38	33·1	1·58 (11), 3·11 (8), 3·73 (30), 4·20 (8), 5·05 (17)	74	3·61
2	0·238	32, 32, 32, 33, 33, 34, 35	33·0	4·85 (13), 5·71 (14), 6·18 (16), 6·33 (15), 7·00 (15)	73	6·03
3	0·077	29, 32, 32, 33, 34, 35, 36	33·0	1·00 (16), 1·10 (10), 1·71 (14), 2·17 (18), 4·31 (16)	74	2·10
4	0·307	31, 33, 33, 33, 34, 35, 35	33·4	1·41 (12), 1·55 (20), 2·86 (14), 4·25 (12), 5·05 (18)	76	3·02
5	0·069	32, 32, 33, 33, 35, 35, 36	33·8	3·23 (13), 3·67 (18), 4·12 (17), 4·83 (12), 5·12 (16)	76	4·18
6	0·123	32, 33, 34, 35, 35, 36, 37	34·4	2·35 (17), 4·29 (7), 5·00 (13), 5·23 (9), 5·34 (18)	64	4·39
7	0·130	31, 33, 33, 35, 36, 36, 37	34·4	4·80 (15), 4·86 (14), 5·50 (10), 6·00 (15), 6·75 (12)	66	5·56
8	0·108	28, 30, 33, 34, 35, 36, 37	33·3	3·10 (10), 5·08 (12), 7·23 (17), 7·69 (16), 8·20 (15)	70	6·42
9	0·130	31, 32, 32, 33, 34, 35, 36	33·3	0·83 (12), 1·28 (18), 2·00 (13), 4·82 (17), 5·64 (14)	74	2·94
10	0·292	30, 30, 32, 32, 33, 34, 35	32·3	7·55 (18), 3·56 (18), 4·00 (21), 6·00 (17), 6·42 (14)	88	4·40
11	0·100	29, 30, 30, 31, 34, 34, 34	31·7	1·37 (8), 2·22 (17), 2·45 (11), 4·17 (13), 7·20 (10)	59	3·45
12	0·184	30, 32, 32, 33, 34, 34, 35	33·0	1·78 (14), 3·18 (11), 4·00 (9), 4·36 (19), 5·35 (20)	73	3·82
13	0·130	31, 32, 33, 33, 34, 36, 38	33·8	5·66 (12), 5·70 (17), 6·62 (16), 8·64 (14), 9·76 (12)	71	7·22
14	0·200	30, 32, 32, 33, 33, 35, 37	33·1	3·92 (13), 4·45 (18), 4·91 (12), 5·33 (21), 6·15 (13)	77	4·92
15	0·184	31, 31, 34, 34, 35, 36, 36	33·9	0·77 (17), 0·82 (11), 0·84 (19), 1·38 (13), 1·47 (34)	94	1·09
16	0·123	32, 33, 34, 35, 36, 36, 37	34·7	0·80 (10), 1·00 (10), 1·18 (28), 1·41 (12), 2·08 (12)	72	1·29
17	0·092	31, 32, 32, 34, 35, 37, 37	34·0	3·46 (13), 4·28 (14), 4·40 (20), 4·68 (16), 5·38 (13)	76	4·44
18	0·085	32, 33, 34, 34, 34, 36, 38	34·4	1·18 (11), 1·30 (10), 2·06 (16), 5·60 (15), 6·43 (14)	66	3·41
19	0·069	29, 33, 34, 35, 35, 35, 37	34·0	0·92 (13), 3·85 (20), 5·21 (14), 5·55 (11), 6·08 (12)	70	4·27
20	0·054	27, 32, 33, 34, 35, 36, 36	33·3	—	—	—

21	—	27, 30, 31, 31, 31, 34, 36	31·4	—	—	—
22	—	No sample	{33·9			
23	—	31, 33, 33, 34, 34, 36, 36	{33·9	—		
24	—	27, 27, 28, 30, 30, 31, 34	29·6	—	—	—
25	—	27, 28, 30, 30, 31, 33, 34	30·4	—	—	—
26	—	26, 28, 30, 31, 31, 32, 35	30·4	—	—	—
27	—	No sample	{31·6			
28	—	28, 28, 32, 32, 32, 33, 36	{31·6		—	—
29	—	29, 30, 32, 33, 34, 35, 36	32·7	—	—	—
30	—	28, 30, 30, 31, 31, 33, 34	31·0	—	—	—
31	—	No sample	{29·7	—	—	—
Nov. 1	—	29, 29, 29, 29, 30, 31, 31	{29·7	—	—	—
2	—	27, 29, 31, 31, 32, 33, 35	31·1	—	—	—
3	—	25, 26, 27, 27, 28, 31, 31	27·4	—	—	—
4	—	25, 26, 30, 31, 31, 32, 32	29·6	—	—	—
5	—	No sample	{29·9	—	—	—
6	—	23, 25, 30, 31, 32, 33, 35	{29·9	—	—	—
7	—	No sample	28·9	—	—	—
8	—	27, 28, 28, 29, 29, 30, 31	28·9	—	—	—
9	—	27, 28, 28, 28, 29, 29, 30	28·4	—	—	—

NOTES TO TABLE III.

Lengths of lint before September 1, and strengths after October 19, were not determined.

The column headed "Number of Bolls Picked per Plant per Day" gives the average number of bolls picked each day *per plant*. The family of plants studied consisted of 130 individuals; thus, 0·100 boll per plant per day implies that every tenth plant had an open boll on it, or that thirteen bolls were picked on that day.

The column headed "Strength of Lint," etc., is given in the arbitrary scale resulting from measurements made on the particular pendulum apparatus employed, as the arithmetical labour of conversion would have served no useful purpose.

The column "Each Bundle tested" gives the number of lint hairs in each bundle in brackets, preceded by a figure showing the mean strength *per hair* in each bundle.

The last two columns give the mean for all the hairs, together with the total number tested. The number given for the mean strength is actually the mean of (*a*) strength of all the hairs individually, and (*b*) all the bundles; in this way any errors due to inequality in bundle size are reduced.

TABLE IV.

DAILY PICKINGS, 1913
(DATA SMOOTHED TO FIVE-DAY MEANS).

Date of Picking.	Lint.		Length by Pulling in Inches.
	Length in Millimetres.	Strength in Arbitrary Scale.	
Aug. 12	—	—	—
,, 13	—	—	—
,, 14	—	4·38	—
,, 15	—	4·22	—
,, 16	—	3·81	—
,, 17	—	3·77	—
,, 18	—	3·25	—
,, 19	—	3·42	—
,, 20	—	3·38	—
,, 21	—	3·51	—
,, 22	—	3·26	—
,, 23	—	3·63	—
,, 24	—	4·09	—
,, 25	—	4·12	—
,, 26	—	4·46	—
,, 27	—	4·60	—
,, 28	—	4·44	—
,, 29	—	4·26	—
,, 30	—	4·95	—
,, 31	—	4·91	—
Sept. 1	—	4·60	—
,, 2	—	4·45	—
,, 3	34·4	4·66	$1\frac{43}{64}$
,, 4	34·2	4·07	,,
,, 5	34·2	3·68	,,
,, 6	34·1	3·69	$1\frac{11}{16}$
,, 7	34·0	3·68	,,
,, 8	34·0	3·49	,,
,, 9	34·1	3·65	$1\frac{43}{64}$
,, 10	33·9	3·87	$1\frac{21}{32}$
,, 11	33·5	4·19	$1\frac{39}{64}$
,, 12	33·3	4·24	$1\frac{5}{8}$
,, 13	33·4	4·09	$1\frac{39}{64}$
,, 14	33·2	3·83	,,
,, 15	33·5	4·12	,,

TABLE IV.—*Continued.*

DAILY PICKINGS, 1913
(DATA SMOOTHED TO FIVE-DAY MEANS).

Date of Picking.	Lint.		Length by Pulling, in Inches.
	Length in Millimetres.	Strength in Arbitrary Scale.	
Sept. 16	33·7	4·03	$1\frac{5}{8}$
,, 17	33·7	3·82	,,
,, 18	34·0	3·76	,,
,, 19	34·1	3·71	,,
,, 20	33·9	3·32	,,
,, 21	34·0	3·48	$1\frac{39}{64}$
,, 22	34·1	4·10	,,
,, 23	34·0	4·12	,,
,, 24	34·0	4·42	,,
,, 25	34·0	4·65	$1\frac{19}{32}$
,, 26	34·0	4·58	$1\frac{19}{32}$
,, 27	33·7	4·58	,,
,, 28	33·3	4·12	,,
,, 29	33·1	3·81	$1\frac{37}{64}$
,, 30	33·0	4·11	$1\frac{9}{16}$
Oct. 1	32·8	3·61	$1\frac{37}{64}$
,, 2	32·9	3·15	,,
,, 3	33·3	3·44	$1\frac{9}{16}$
,, 4	33·5	3·94	,,
,, 5	33·8	3·85	$1\frac{19}{32}$
,, 6	33·9	4·71	$1\frac{39}{64}$
,, 7	33·8	4·70	$1\frac{19}{32}$
,, 8	33·5	4·74	$1\frac{39}{64}$
,, 9	33·0	4·55	,,
,, 10	32·7	4·20	$1\frac{19}{32}$
,, 11	32·8	4·36	$1\frac{39}{64}$
,, 12	32·8	4·76	$1\frac{39}{64}$
,, 13	33·1	4·10	$1\frac{19}{32}$
,, 14	33·7	3·67	$1\frac{9}{16}$
,, 15	33·9	3·79	,,
,, 16	34·0	3·03	$1\frac{35}{64}$
,, 17	34·2	2·90	$1\frac{17}{32}$
,, 18	34·1	—	—
,, 19	33·4	—	—
,, 20	33·4	—	—

TABLE IV.—*Continued.*

DAILY PICKINGS, 1913
(DATA SMOOTHED TO FIVE-DAY MEANS).

Date of Picking.	Lint.		Length by Pulling, in Inches.
	Length in Millimetres.	Strength in Arbitrary Scale.	
Oct. 21	32·3	—	—
,, 22	32·4	—	—
,, 23	31·8	—	—
,, 24	31·6	—	—
,, 25	31·2	—	—
,, 26	30·7	—	—
,, 27	31·3	—	—
,, 28	31·5	—	—
,, 29	31·3	—	—
,, 30	30·9	—	—
,, 31	30·8	—	—
Nov. 1	29·9	—	—
,, 2	29·5	—	—
,, 3	29·5	—	—
,, 4	29·6	—	—
,, 5	29·1	—	—
,, 6	29·4	—	—
,, 7	29·2	—	—
,, 8	—	—	—
,, 9	—	—	—

NOTES TO TABLE IV.

The columns for "Length" and "Strength" are obtained in the same way as in Table II.

The column headed "Length by Pulling, in Inches" has been added to give a basis of comparison with the ordinary expression of lint length, and also to show the unreliability of length determinations made by "pulling," as compared with combing the seed-cotton. This determination was made on each sample from September 1 to October 20.

LIST OF REFERENCES

THE following list is supplementary to the one published in "The Cotton Plant in Egypt," London, 1912. Works included in both lists are marked with an asterisk.

*BAINES, EDWARD: History of the Cotton Manufacture. London, 1834.

BALLS, W. LAWRENCE:

33. The Cotton Plant in Egypt: Studies in Physiology and Genetics: *Macmillan's Science Monographs*. London, 1912. With bibliography, 1-32.
34. Seed Breeding and Deterioration of Cotton Varieties. Two Lectures delivered to the Egyptian Congress of the International Federation of Master Cotton Spinners: *Report of Congress*. Manchester, 1913.
35. Electrical Response in Cotton Plants: *Annals of Botany*, vol. xxvii., 1913.
36. Meteorological Conditions in a Field Crop: *Quarterly Journal of the Royal Meteorological Society*, vol. xxxix., April, 1913.
37. The Movements of Soil-Water in an Egyptian Cotton-Field: *Journal of Agricultural Science*, October 7, 1913.
38. A Study of Some Water-Tables at Giza: *Cairo Scientific Journal*, May, 1914.
39. The Sudan and Cotton: *The Near East*, April and May, 1914.
40. Leaf-Fall as a Factor in Soil-Deterioration. Proceedings of Cambs Philosophical Society, May, 1914.
41. Specific Salinity in the Cell-Sap of Pure Strains. *Ibid.*
42. Pre-Determination of Fluctuation: A Preliminary Note. *Ibid.*
43. Grafted Cotton: *Worrall's Textile Monthly Gazette*, August, 1914.
44. Science and the Supply of Fine Cotton: *Science Progress*. October, 1914.

45. A Central Department of Cotton Affairs: *Worrall's Monthly Textile Gazette*, November and December, 1914.

46. Evading the Boll-Worm: *The Near East*, December, 1914.

In collaboration with F. S. HOLTON.

47. Analyses of Agricultural Yield—Part I., The Spacing Experiment with Egyptian Cotton, 1912. Transactions of the Royal Society, 1915.

BARKER, A. F.: Textiles. London, 1910.

BARWICK, F. W.: Defects in Fabrics. Paper read before the Textile Society of Manchester School of Technology, April, 1913

BEECH, FRANKLIN: The Dyeing of Cotton Fabrics. London, 1901.

BERSCH, J.: Cellulose, Cellulose Products, and Artificial Rubber. Translated by W. T. Brammt. Philadelphia and London, 1904.

*BOWMAN, F. H.: The Structure of the Cotton Fibre. Manchester, 1881; second edition, London, 1908.

BROWN, SIR HANBURY: Irrigation. Second edition. London, 1912.

*COOK, MCLACHLAN, AND MEADE: A Study of Diversity in Egyptian Cotton. U.S. Department of Agriculture, Bur. Pl. Ind., Bulletin 56, July, 1909.

COOK, W. H.: The Cotton Industry. Chapter in Textiles by A. F. Barker, *q.v.*

CROSS AND BEVAN: Researches on Cellulose. London, 1900 *et seq.*

DOBSON, B. PALIN: The Story of the Evolution of the Spinning-Machine. Manchester, 1910.

FLATTERS, A.: The Cotton Plant, its Development and Structure; and the Evolution and Structure of the Cotton Fibre. Manchester, 1906.

*HANNAN, W. I.: The Textile Fibres of Commerce. London, 1902.

*HENRY, YVES: Determination de la Valeur Commerciale des Fibres du Coton. Bull. du Jardin Colonial et des Jardins d'essai des Colonies, 1902.

KEARNEY, T. H.: Mutation in Egyptian Cotton: *Journal of Agricultural Research*, July, 1914.

LAMBORN, L. L.: Cotton Seed Products. New York and London, 1904.

LEIGH, EVAN: The Science of Modern Cotton Spinning, 2 vols. Manchester, 1873.

LESTER, J. H.:

1. The Hygroscopic Nature of Cotton: *Journal of the Society of Chemical Industry*, March, 1902.
2. Moisture in Cotton and Yarn. Paper read before the British Association of Managers of Textile Works. Manchester, 1904.
3. Testing of Textile Material. *Ibid.* Manchester, 1909.
4. Scientific Instruments for Testing Textiles. Annual Congress of the Textile Institute. Manchester, 1910.

LLOYD, F. E.: Leaf Water and Stomatal Movement in Gossypium: *Bulletin of Torrey Botanical Club*, January, 1913.

MATTHEWS, J. M : Textile Fibres, pp. 110-169. New York and London, 1904. Especially p. 159, footnote.

MCCONNEL, J. W.: Some Thoughts on the Introduction of New Cottons. *Textile Mercury*, March, 1914.

Commerce and Science in Cotton Growing. Third Int. Conf. Tropical Agric.; *Report*, London, June, 1914.

MONIE, HUGH:

The Cotton Fibre, and the Mixing of Cotton. London, 1904.

*The Cotton Fibre, its Structure, etc. Manchester and London, 1890.

MORRIS, J., AND WILKINSON, F.: The Elements of Cotton Spinning. London, 1897.

O'NEILL, C.:

1. On an Apparatus for Measuring Tensile Strengths, especially of Fibres. *Mem. Lit. and Phil. Soc. Manchester*, November 17, 1863.
2. Experiments and Observations upon Cotton. *Ibid.*

POSSELT, E. A.: Cotton Manufacturing, 2 vols. London, 1903.

SCHANZE, MORITZ: Cotton in Egypt and the Anglo-Egyptian Sudan. Ninth International Congress of the International Federation of Master Cotton Spinners at Scheveningen, July, 1913.

SCHMIDT, ARNO: Cotton Growing in Egypt. International Federation of Master Cotton Spinners Egyptian Congress Report. Manchester, 1913.

TAGGART, W. SCOTT: Cotton Spinning, 3 vols. London, 1906-08.

TODD, JOHN A.:

1. The World's Cotton Crops. London, 1915.
2. *The Market for Egyptian Cotton in 1909-1910: *L'Égypte Contemporaine*, vol. i., p. 1.
3. *Further Notes, 1910-1911: *Cairo Scientific Journal*, January, 1912.

UTTLEY, T. W.: Cotton Spinning and Manufacturing in the United States of America. Manchester, 1905.

*WATT, SIR G.: The Wild and Cultivated Cottons of the World. London, 1907.

*WEISNER, J.: Die Rohsteffe des Pflanzenreiches: B. ii., Fasern Samen. Leipzig, 1903.

*WILLCOCKS, SIR WILLIAM: Egyptian Irrigation, third edition. London, 1913.

ZIPSER, JULIUS: Textile Raw Materials, translated by Charles Salter. London, 1901.

INDEX

BILLING AND SONS, LTD., PRINTERS, GUILDFORD, ENGLAND.

Made in the USA
Monee, IL
13 September 2026

62976741R00163